Ephemeral Hunter-Gatherer Archaeological Sites

Ephemeral Hunter-Gatherer Archaeological Sites

Geophysical Research

Jason Thompson
School of Advanced Studies, Center for Educational and Instructional
Technology Research, University of Phoenix, Phoenix AZ, USA

ELSEVIER

AMSTERDAM • BOSTON • HEIDELBERG • LONDON
NEW YORK • OXFORD • PARIS • SAN DIEGO
SAN FRANCISCO • SINGAPORE • SYDNEY • TOKYO

Elsevier
Radarweg 29, PO Box 211, 1000 AE Amsterdam, Netherlands
The Boulevard, Langford Lane, Kidlington, Oxford OX5 1GB, United Kingdom
50 Hampshire Street, 5th Floor, Cambridge, MA 02139, United States

British Library Cataloguing-in-Publication Data
A catalogue record for this book is available from the British Library.

Library of Congress Cataloging-in-Publication Data
A catalog record for this book is available from the Library of Congress.

ISBN: 978-0-12-804442-1

For Information on all Elsevier publications
visit our website at https://www.elsevier.com/

Working together
to grow libraries in
developing countries

www.elsevier.com • www.bookaid.org

Publisher: Candice Janco
Acquisition Editor: Marisa LaFleur
Editorial Project Manager: Marisa LaFleur
Production Project Manager: Paul Prasad Chandramohan
Designer: MPS

Typeset by MPS Limited, Chennai, India

Cover illustration: The 13 March 2005 explosion and pyroclastic flow at Volcán de Colima,
México with associated seismic signal (Photograph of Sergio Velasco García).

CONTENTS

PREFACE

What does it mean to be an archaeologist, or a geophysicist, or both? Those readers who know me will already be familiar with my use of rhetorical questions. I will pose many of them in this book. Some of them will even be answered. Others will be left hanging, hopefully picked up, to be answered in future dialogs. This is a book intended to stimulate discussions, not to end them. I certainly don't have all the answers or even most of them, nor would I want to have them. All answer and no question leaves Archaeologist, in my opinion, a very dull companion. I certainly don't want to write such a book, and it is doubtful anyone would want to read one like that. This book is itself broadly about archaeological geophysics (AGP), often termed near-surface geophysics, or even geophysical archaeology (GA) (Conyers, 2013; Thompson et al., 2011). But this is not a How-To book. There are already very many of those in the literature. This is a book about taking chances and pushing envelopes, devoted to the development of method and theory suitable to the use of remote AGP techniques *as a mechanism or platform of observation:* observation of human phenomena and human behavioral conditioning of the archaeological record, specifically, instead of just as a way of finding or not finding things and places to excavate. How many of the AGP pros use it *explicitly* from an observational perspective to investigate human behavioral phenomena? If I could have exactly what I want, it would be an archaeoethnographic AGP, one where AGP is employed along with other methods in a battery of techniques to make observations about and to draw out data from the human past. This AGP would function through appeal to a paradigm such that AGP is serviceable as a human-observational platform. It is not merely a set of gadgets to find stuff in the ground or areas of the subsurface devoid of interesting things. My ideal AGP would be one where dirt and lab archaeologists are also involved in formulating AGP research and publishing protocols instead of being regarded as uninvolved funding streams, customers outside the inner circle who either won't, can't, or aren't allowed to provide input.

AGP has a lot of currency these days, even though it is hardly ubiquitous in American academic or CRM archaeology. Yet, despite its growing presence in the archaeology mainstream, I suspect there are few if any degrees granted in GA/AGP, few American practitioners permanently devoted to its exclusive use active in the United States, and few Anthropology departments equipped to train students either to be adept or (and here is the rub) *critical* of it. If not many people really understand, they certainly are in no position to criticize it, and this is also a purpose of this book: to draw interested candidates into the fold. Pretend that digital calipers were exclusively handmade and extremely cost-prohibitive as is the case with geophysical equipment. Imagine further that faunal analyses using such devices were regarded as though they were some mystic art known only to a few initiates who possessed the rare and mysterious digital calipers.[1] How many of us would be in a position to offer cogent criticism of it or even to make use of its methods?

The use of AGP will only become more prevalent in the future and should, therefore, be an integral part of American archaeology undergraduate and graduate programs. Yet, instruction in AGP techniques is at present available only to a minority of American archaeology students. Most students who wish to acquaint themselves with AGP have to take courses outside their enrolled major programs of study (if they are fortunate enough to attend universities that own near-surface geophysical equipment accessible to students), or worse, attend extremely expensive off-campus GA courses with high overhead travel costs offered by the lucky few archaeologists and institutions with access to the equipment. For another rhetorical question: what if the use of digital calipers was similarly restricted to mostly zooarchaeologists, physical anthropologists, and their students? I doubt those scenarios would be regarded warmly. AGP, similarly, should be something open and available to interested people, not controlled and administered preciously.

Another primary reason I am interested in this topic regards the relative frequency of archaeological sites; or, rather, the decreasing frequency of them relative to the increasing global human population.

---

[1]These are themes to which I will return often, out of curiosity, self-interest, and scholarly focus. I think the entrenched structural norms of AGP research and the related peer review process deserve some special scrutiny for reasons that I will elucidate.

More people probably means, among many other things, more archae-ologists competing for access to fewer sites. Along with decreasing public investment in archaeology and Science in general, this places archaeology students in the unenviable position of pursuing a major in a discipline for which primary data and excavation experience can be expected to be problematic. We might even regard archaeological sites as an endangered species, one which cannot be managed into a situation of artificial plenitude by virtue of salvage breeding. Archaeological sites do not reproduce. Anything archaeologists can do to preserve known sites without destroying them through excavation is important, and this involves AGP as perhaps the best means of testing sites noninvasively (Conyers, 2012, 2013; Thompson, 2015). I wonder, however, if the current "prospecting" model of AGP application is really suited to learning about most sites in general and especially small-scale, open-air hunter-gatherer sites (SSS) in particular. SSSs have certainly not figured as predominantly as megasites. That appears to be virtually a built-in focus for much AGP publishing. As in, "we took it over there, we used it like that, here are the pictures we made, where's the next job?"

Since I asked what it means to be an archaeologist above, let's now explore what it means to be a geophysical archaeologist (GA). What established Anthropology or Archaeology curricula do GAs learn? Are they trained to recognize anthropogenic lithics and faunal material from background noise? How does their training articulate with Anthropology? One could also ask whether a familiarity with AGP technology and the ability to use it in the field are necessary and suffi-cient qualifications for lucrative employment as an Archaeologist or Anthropologist. It seems to me that, at least for some GAs (Conyers, 2010, 2012, 2013; Conyers and Leckebusch, 2010; Thompson, 2014, 2015; Thompson et al., 2011), such AGP familiarity and technical abilities not only qualify one to self-identify as an archaeologist; it actually appears that AGP equipment savvy, prospecting techniques, and especially the construction of a literary presence based on them *are becoming or have become by themselves a specific type of archaeol-ogy in its own right to a small but entrenched GA cadre.* Explanations and understandings of actual people, using AGP as a *mechanism of observation*, of the various materials they behaviorally modified and conditioned in the past, are apparently mundane to judge from within the AGP primary literature. More text in AGP is devoted, for instance, to descriptions of equipment settings and wave parameters than to anything or anyone they are generally used to locate. Imagine if dirt archaeologists were disposed to discuss trowels this way.

I will admit to being an anthropological archaeologist (AA) who wants to use AGP to study human behavior in a "new and different" (or, "new and improved") way. If in the course of studying behavior I might happen to locate a new site or unknown parts of a known site, great. Does a cursory, jargonized, and superficial description of remote AGP images of buried material situated within vast treatises of pure Physics equate to studying the archaeological record? Many GAs publish as though this were the case. It seems, at least to me, that AGP is often used more frequently to analyze the equipment and the physics and reflective properties of materials than it is used to study humans and past human behavior. If my goals and objectives, nebulously as they have been expressed so far, seem naïve then perhaps they are and perhaps I am. So be it.

Why do AAs study the archaeological record? Imagine a scenario in which, say, Francois Bordes himself argued that minute descriptions of the physical dimensions and the material used to make a lithic flake—only the material and its physical properties (as opposed to the flake itself, as an artifact of past human behavior), the whole physical properties and nothing but the physical properties—*and especially the magnifying glass used to see them and the prose used to describe the seeing*—was not merely archaeology but was in fact the entire point of it. Would that be Archaeology? Is that a human science or an optical one? If you are skeptical about that satire, then I simply suggest you read several examples in the dominant archaeology journals or books. I hope you like reading about really big things like castles and pyramids and their relative dimensions buried inside immense overburdens of sediment, jargon, and math. People? Culture? Behavior? Those things don't seem to have nearly as much currency in those venues. Because it seems that AGP has been, like many famous actresses and actors, "typecast." It refers to the formulaic.

This will also be a book dedicated to the ignoble small-scale ephemeral archaeological site and some of the human behaviors that formed them, along with geophysicsm, used to illustrate my points as opposed to being the focal content. Those of my colleagues who know me personally will probably find this rather a natural pursuit for me.

Those who don't or haven't read my previous publications will probably be new to the subjects I will discuss and especially the manner in which I relate them. To be blunt, I am trying to infuse AGP with anthropological research premises that it currently and in large measure lacks.

Upon further reflection, many of my GA and other AA colleagues just don't get it. For example: try explaining to your paleoanthropologist or Paleolithic archaeology familiars that you are working out a book on the application of AGP specifically and only to small-scale ephemeral sites (SSSs). One gets puzzled looks and responses from the few who bother to reply at all. You see, AGP is like a very small shire with a very interested posse of shire-reeves who think it should be this or that within a very small range of variation, and they don't really care what AAs think because it's their *cosa nostra*. I want it to be much more than it is or than the reeves think it should be, and I suspect some others do as well, which is another reason for this book. AGP could be a lot more with new thinking applied to it. Also, the fact that practically anyone can be trained to use the equipment suggests that archaeologists can learn how to do this all themselves and cut out middlemen.

As it seems to me, despite the relative progress AGP has made recently in approaching questions and issues of theoretical and methodological significance to Anthropology instead of pure physics, many matters are still left unaddressed in the relevant literature. Ask 10 archaeologists what AGP is and you will get a variety of answers, many symptomatic of the topical biases inherent in the primary literature itself. Variations on the following theme are extremely common: "Oh that stuff is for castles and forts and such. I don't see how that can apply to ephemeral open-air sites." We can be forgiven for such ignorance when each successive year new publications describe exciting new possibilities for using AGP at the very same or at least the same kinds of large-scale sites with visible surficial components that were printed in the last volume. Pyramids, temples, castles, walls, bridges, etc., do tend to give rather a stark indication that probable subsurface material appurtenances exist. Ephemeral Paleolithic or Paleoindian sites are less obvious. Certainly other people noticed this wide open literary niche, most probably well before I became aware of it. Yet, strangely few people seem to be publishing about it, or perhaps few are getting their research past an entrenched AGP peer review process/ cohort.[1] So I guess that's the main reason for this book: nobody else wanted to write it or saw reason to write it or got it into review. And away we go.

CHAPTER 1

Archaeological Geophysics Biases

Sometimes books and articles seem almost to write themselves. This is not one of those. This book will be focused instead upon the author's own thinking, some assorted rougher ideas, and methods for using archaeological geophysics (AGP) and combining them with other, more established archaeological techniques, exclusively at open-air small-scale ephemeral archaeological sites (OSSSs or small-scale sites of any interval, SSSs). The intended audience is anyone who might be interested in using AGP techniques at archaeological sites, from post-PhD professionals to CRM archaeologists, and especially archaeology students at both undergraduate and graduate levels. Why OSSSs as an exclusive focus? Although the rest of the book will take up the challenge of elucidating that theme in depth, a brief introduction to some basic facts about AGP should suffice as a preliminary explanation. First, although AGP is not a "new" archaeological technique, its widespread application in the United States is new, at least in comparison to its use in Eurasia (Conyers, 2012, 2013; Thompson, 2011, 2014, 2015). Second, in response to a call for a more anthropologically oriented AGP from several of its more visible practitioners (Conyers, 2010; Conyers and Leckebusch, 2010), one has taken that particular challenge seriously and has attempted to assemble a coherent research strategy in response (Thompson, 2011, 2014, 2015), of which this book is a part. Lastly, some extreme publication biases are quite prevalent in AGP research and publishing, and these biases extend beyond particular scholars, publication venues, and continents (Thompson, 2014, 2015) and one suspects this heavily colors the reception AGP gets from "anthropological audiences."

What is an "anthropological audience"? By that phrase, one means merely a range of scholars who are attuned to the theory and method of Anthropology, as taught and practiced in four-field Anthropology departments as the holistic study of humankind. Not all archaeologists consider themselves to be anthropologists; most anthropologists would

Ephemeral Hunter-Gatherer Archaeological Sites. DOI: http://dx.doi.org/10.1016/B978-0-12-804442-1.00001-9

likewise not define themselves as archaeologists. Yet, in the United States anyway, it is extremely common for students of Anthropology to receive undergraduate and graduate instruction and training in all four of the following subdisciplines (in no particular order): Sociocultural Anthropology, Linguistic Anthropology, Archaeology, and Physical or Biological Anthropology. However, one can search many AGP journal articles and books in vain for any mention of Anthropology, and that is a problem. Specifically, the problem relates to AGP's inclusion in US Anthropology and especially US Archaeology departments, most of which assume that faculty and students both subscribe to an understanding of Anthropology as the holistic study of humans. For while AGP can be used to address certain issues of archaeological significance, even a cursory review of AGP publications reveals that they are often only tangentially related at all to Anthropology and hardly ever premised upon the study of humans in any way. Moreover the community of established AGP practitioners, or at least those who publish frequently and review peer publication submissions and grant proposals, appears to have members who lack an interest in addressing humanistic anthropological themes and perhaps even resents attempts at such scholarship from perceived outsiders (Thompson, 2015).

A reluctance to accept the input from nongeophysical archaeologists could result from many factors, not the least of which is simple momentum. Established wheels don't always need reinvention. Yet, a comfort with established procedures and habits in contemporary AGP publication and research norms has resulted, at least thus far, in an imbalanced literature. One common pattern is as follows: A researcher or a research team uses either one AGP technique or an array of them to test the feasibility of application at a known archaeological or historical site with enormous surface structures present, perhaps to acquire data regarding buried appurtenances of the surface material. Pictures are generated from the acquired data. The study is published and, all too often, repeated at another site similar to the first one in most respects, or new and improved data collection techniques using one or more AGP methods is conducted at the same site. Moreover the sites themselves in this literature all seem to cluster oddly into the late Bronze Age through Classical and Medieval intervals: if Greek, Roman, Nabataean, and Medieval Eurasian megasites were forbidden from publication what would remain of the AGP literature? A focus on monumental Egyptian and Mesoamerican sites is also evident, the connection being the overtly redundant attention to massive materials. Imagine if zooarchaeologists only typically analyzed the remains of mammoths, mastodons, and other assorted Pleistocene megafauna such as megaceros and rhinoceros, and virtually excluded analyses of microfauna from discussion. How reliable a portrait of the past would that paint?

Perhaps huge sites can inform us about huge human sociocultural processes, at least to some extent. The existence of roads, walls, foundations, etc., at a pyramid or castle site obviously provides information in the aggregate about the former presence of human institutions devoted to building pyramids, roads, walls, foundations, etc., just as we can observe in our own societies today. Since we can observe large-scale construction activities today, it does not require an enormous imaginative leap to retroject similar behaviors into the past to explain the existence of ancient pyramids, roads, walls, foundations, etc. But what else do such building programs do? How do they impact the archaeological record? Why do we today have laws regarding the protection of cultural resources from impact by construction projects? Such projects destroy the archaeological record where it exists and obliterate material signatures of nonarchaeological human behavior apart from that. This brings us to another reason for an emphasis on SSS: SSSs offer us the opportunity to learn about the behaviors of individuals and small groups instead of entire societies and the economics or politics of state societies. A herd obliterates the footprints of the individual.

Obviously the archaeology of the individual and small groups pertains to Large-scale sites (LSS), it is just that their actual material signatures in the archaeological record were often obliterated by construction activities or immersed within indecipherable palimpsests. LSS themselves frequently represents numerous successive modifications and remodifications of landscapes and environments, whereas SSS portray adaptations to them; these are very different things. Additionally, such petite data from LSS are not just invisible to AGP; they are invisible to most or all other archaeologists as well, having been erased by successions of human activities. Although this is speculative, it likely does not require much specialized training to identify large or monumental structures, streets, walls, public spaces, etc., even in images of AGP data. One can simply observe many examples of similar things in our own lives. In these ways, analysis of SSS involves articulations between anthropological method *and* theory as opposed to being primarily methodological application or reapplication exercises as it is with giant things.

Although the contemporary AGP literature represents much methodological innovation and improvements in practical applications to archaeological prospecting, has it advanced thinking in Anthropology or Archaeology? Most refinement in AGP appears to present methodological (generally to increase imaging resolution or clarity) as opposed to theoretical development. Thompson et al. (2011) provided some framework for how to position AGP in relation to Archaeology by posing research questions, and yet the AGP literature is otherwise rather deficient in terms of formulating such interrogatories. For example, how can AGP be used to infer variable seasonality of archaeological occupations, or differential residential mobility, labor organization, occupying cohort size, differential animal carcass-processing and depositional disposition, or even perhaps relative occupational duration at sites (some of these will actually be answered below!)? As indicated in the introduction, rhetorical questions figure largely in this book, mainly to stimulate further questions and means of answering them.

Is contemporary AGP publishing premised upon the stylistic repetition of feasibility studies looking at known megastructural sites and looking for buried portions of them (prospecting)? As a Paleolithic Archaeologist, one finds this and other structural biases inherent in AGP publication puzzling. For perhaps more than 99% of the time humans have existed on this planet they lived as hunter-gatherers (h/g), which suggests that SSS ought to outnumber LSS (even considering post-Neolithic human population growth), and yet why the overwhelming bias toward AGP applications on megastructural, post-Neolithic material? If there exists a widespread archaeological bias for AGP to be used primarily as a prospecting technique, then is no one prospecting for SSS? If not, why not? Is there an unanticipated or unexplained shortage of Paleolithic, Mesolithic, or North American Paleoindian or Archaic sites in relation to castles and pyramids? When such SSSs are conspicuously absent from AGP publications, it becomes even more interesting to observe the fact that most of the nongeophysical, primary peer-reviewed archaeological journals haven't noted this rarity. *American Antiquity*, for just one example, publishes many articles on archaeology at SSS, and has explicitly focused upon themes of anthropological theoretical contributions as opposed to site-spotting.

Something *American Antiquity* does not seem to publish is formulaic recapitulations of faunal analyses, for instance, of faunal assemblages that have already been published there or elsewhere. If SSSs are truly as rare as AGP publication norms suggest, then wouldn't it follow that rare AGP research at them should be valued by their larger academic community instead of a perhaps too-frequent reception as unwanted visitors?[1] Finally, whereas Anthropology and Archaeology have grappled with financial and gender equity in teaching and scholarship, is there no visible bias in AGP research conducted on the structural remains of megastructures built by commission of dead one-percenters?

1.1 SPECIAL NOTES TO ARCHAEOLOGY STUDENTS

Not to worry, I haven't forgotten that this book is for you too. Welcome to Archaeology. Unquestionably, at this point in your scholarly tenure you have been received by concerned parents and other family members, most of them aghast at the field you have chosen as a potential occupation. This is quite normal, you should know. If anyone has gotten uncouthly wealthy by doing archaeology, then they are keeping very quiet about it. Archaeology is an often brutal, thankless "job" that hardly anyone outside our small community understands or cares about, serving as the frequent subject of many jokes and cartoons (Gary Larson and *The Far Side* come to mind). If you really want to "do" this, then just know that you are likely to be consigning yourself, at least for considerable periods, to a life of economic uncertainty, endless job searches, people who think you work on dinosaurs or who ask what you think of the last "mysterious" archaeology TV program they saw (those shows are *always* presented as mysteries...TV people seem to think we don't know anything about the past even though they keep asking us about it), and many other ignominies. If you make it through an undergraduate program intending to pursue Archaeology further, congratulations and take heed: you now face *at least* two to three further years of

---

[1] The present author submitted his contributions to the research above in book-length form to an alternate publisher outside the normal AGP venues, received a highly enthusiastic reception from the editors, helpful, professional peer review from the reviewers selected by the publisher, and promptly published it.

study in a Master's program before you can certify yourself as a Registered Professional Archaeologist (RPA).

In an archaeology graduate program, you can gleefully anticipate what we, in the grift, colloquially refer to as "seminar," where you read hundreds of pages of technical articles and book excerpts, normally in really poor photocopies, meet once a week, and take turns "presenting," by which we mean taking turns summarizing and synopsizing the endless assigned readings to the assembled class cohort. Unfortunately for all of us, Anthropology is today moving so fast and overturning so many prior assumptions that by the time many things are published they are already obsolete. We used to think a lot of things, such as how Neanderthals were a "species" reproductively isolated from "modern humans," and that archaic "species" like the Neanderthals lacked any symbolic abilities, imagination, or language whatsoever. Such orthodoxies served the disciplinary, literary, and certainly the financial interests of the very people who formulated and propagated such Anthropological orthodoxy. The recent media coverage of the complex Neanderthal structures built at Bruniquel Cave have suddenly and further obliterated our precious orthodox Anthropology view of our "archaic" European ancestors: they built stone structures well before "modern" humans existed (*Nature* 2016 doi:10.1038/nature18291). This will have the effect of further media coverage, where reporters and bloggers request the input of "high-stature" Anthropologists, who will carefully explain that "we just don't know enough yet" about Neanderthals but that they are working on it. "I was wrong," is something you will have to look long and hard to see one of us says on the record.

A graduate archaeology student in a devoted archaeology program will perhaps complete three or four seminars as well as other more particularized items in your own focused program of study. A graduate archaeology student in an Anthropology program will also need to take seminars in the other three subdisciplines. You could read more than you ever wanted about Pierre Bourdieu and *habitus*. Other topics might include population genetics and the intricacies of primate behavior and ecology, speech communities, signaling theory, and the frightening real-world phenomenon of rapidly decreasing cultural and linguistic diversity. You are obviously expected to be virtually an instant expert at all of this material: you can bet one or two of your class colleagues will certainly seem to be, and it is doubtful you will ever see them outside of class. After that? If you've had enough, you collect your MA or MS and you can fill out some forms, pay a fee, and you will be an RPA. You could try to get a job at a CRM firm, but beware that they are flooded with applicants. If you are truly a masochist, you will continue into a PhD program and apply with hundreds or thousands of people just like you for tenured jobs.

As you while away the years in graduate school, you will also likely find yourself becoming inured somewhat to the theory and practice of "scholarly critique." This is what we call the process where each of your ideas is (theoretically, but hardly ever practically) shredded mercilessly and your thoughts and feelings (sanity too) are reduced and "deconstructed" until they scarcely resemble what you initially thought or wrote. This is by design, because when you begin submitting your ideas to the publication process you will learn about the unvarnished harshness of peer review. Some of the comments you will receive will be thoughtful, cordial, and given in the spirit of helping you to polish your research into a publishable form. Yet, be advised that much of the critique offered you will seem as though you have written something abysmally stupid. Although we don't mention it as often as we should, the peer review process is often turned on its head and perverted into a kind of competitive intelligence program. Some seemingly negative reviewers will dismiss your work and reject it for publication *as they take notes of your topics, methods, and research designs.* Don't be surprised if you later see bits of your own ideas revamped and *published* in somebody else's paper. This happens. You never know when a casual utterance about your ideas will emerge even in a trusted and respected colleague's work. Until you publish it, it isn't yours, and even after you publish it, it actually belongs to the publisher. Why does one mention this in a book about AGP? Here's why: if publishing in Anthropology or archaeology is somewhat cut-throat and often a severe exercise, attempting to publish in AGP venues as an "outsider" can seem personally painful, as you shall see.

1.2 INTERESTING ASPECTS OF "ARCHAEOLOGICAL" PUBLICATION

Imagine the following satirical fictional scenario: a faunal analyst/ zooarchaeologist (the terms are often used interchangeably) conducted an analysis of a given faunal assemblage from Site X. Let us pretend further that instead of premising their study upon an appreciable aspect of the human behavior that conditioned the faunal assemblage, the zooarchaeologist chose instead to focus purely upon physical properties of the bones in their sample. In our case at hand, the analyst decided to frame their analysis on the minute recording of cortical bone thickness among a sample of diaphyseal fragments using several brands of digital calipers in order to test the comparative accuracy of the instruments. Once the study was completed, the analyst compiled clear and informative data tables, ran some statistical procedures comparatively to estimate caliper accuracy, and wrote up the research in a tight, succinct professional manuscript. Let us assume the analyst presented the final paper for publication to *American Antiquity*, for example. For the record, recall that this structurally excellent manuscript with magnificent grammar and vocabulary contained utterly no references to human behavior, culture or specific cultures, or even humans themselves: it was restricted to measurements of cortical bone thickness to test digital calipers.

What would we assume to be the peer review reception of this paper submitted to *American Antiquity*? Although it is impossible actually to know for sure how the preceding fictional project would be reviewed, it is likely that the absence of anthropological content and research framing would be brought to the scholar's attention. Our fictional writer would be advised to frame the research in a manner adequate to address anthropological and archaeological theories and methods and perhaps to focus on a particular time interval or site. The paper would probably be rejected with recommendations for revision and resubmission. It is also likely that at least one or even several of the reviewers might be interested in the intercomparability of different brands of digital calipers, having professional or other interests in instrument accuracy, and would express it. Although one or more reviewers might suggest submission to alternate scholarly journals, it is unlikely that the writer's entire project would be dismissed and rejected utterly and contemptuously even though it featured no anthropological merit in the initial peer review. It is also likely that both faunal specialists and nonspecialists alike reviewed the paper.

Now consider the following not-so-fictional or -satirical scenario: a team of international archaeologists conducted an exhaustive AGP application at Upper Paleolithic Site Y where extensive prior excavations had occurred over decades. Site Y was known to contain unexcavated archaeological material in multiple different areas; in fact, photographs of wall exposures verified the presence of this unsampled archaeology in situ. The unexcavated material at Site Y had known spatial coordinates accurate to the centimeter both laterally and horizontally through the use of a laser theodolite in mapping. The study at hand was premised upon a suite of theoretical and methodological research objectives framed *explicitly from an anthropological awareness*, including the ability of AGP to relocate (from the most recent previous excavations) known but unexcavated anthropogenic deposits with specific coordinates accurate to the centimeter laterally and horizontally. This was done for estimating: site component size in unexcavated portions; the relative number of domestic features at the site; a rough estimate of the number of human individuals necessary to assemble the site; and the development of user methods for discerning between AGP data of actual archaeological material and data from false-positive subsurface matrix displacement (i.e., disturbances from prior excavations and backfilling episodes). The authors successfully relocated the known unexcavated material purely with AGP, compiled voluminous AGP data, composed clear AGP graphic images from the data, invented a method to distinguish between real and false archaeological materials purely from the AGP data, ground-verified the unexcavated material (i.e., excavated the areas AGP indicated to contain archaeological deposits at a later date) and wrote up the research in multiple tight, succinct professional manuscripts. What would we assume to be the peer review reception of these real papers submitted to a venue to which I will just refer as The Leading AGP Journal?

In this case, no commentary in the peer review was devoted to any appreciation for the anthropological merit of the study and the intricate research design (including perhaps the most exhaustive ground-penetrating radar (GPR) survey of any piece of land on the planet at multiple analytical scales and resolutions). In the case above (and in several subsequent peer reviews based upon it), the authors were accused of "not getting it," or worse, of conducting unimportant and trivial research. One recalls "pointless" being one adjective used throughout one reviewer's comments. The project team was also criticized for a perceived failure to include more detailed descriptions (...descriptions...not analysis, epistemology, logic, or even writing ability, but mere descriptions) of equipment settings. Descriptions of such equipment were certainly not lacking but they were criticized for failing to address purely geophysical concerns *in a manuscript premised upon the use of AGP to address issues of anthropological interest not geophysical purity.* Given the tenor of these reviews, one suspects that few *if any* non-AGP purists were consulted for the review, indicating another aspect of AGP publication bias: only a very small population of practitioners is held to be qualified to review or conduct such research. As a final observation, one must state that the team involved in the latter case above included two senior scholars of impeccable repute and extensive publication records and a classical archaeologist with over a decade of experience on several continents using AGP in their own research.

1.3 THE *COSA NOSTRA* ATTITUDE

Why then, you might ask, is another book from this writer necessary? *Precisely because nobody else appears to be writing it.* Apart from the structural biases in the AGP literature addressed previously, it is merely factual to note that many if not most of the primary sources and journals devoted to publishing AGP research are dominated mainly by European scholars who publish papers on mainly Eurasian archaeological sites primarily describing applications at post-Neolithic megasites with obvious surficial expressions that characterize or inform about subsurface material. A professor once said in an undergraduate Roman Archaeology course: "The Romans basically considered the Mediterranean Sea as theirs. The Romans called the Med Our Sea, or 'our lake', or *mare nostrum*" (Storey, 1997, personal communication). The mafia called their pursuits *"cosa nostra"*...Our Thing It seems that some AGP experts might consider AGP to be their own special domain. Although academic peer review is certainly often ruthless and bloody, it is also not normally demeaning, unprofessional, ad hominem, and desultory in tone or spirit. We develop thick skins in this discipline. Nor are all reviews negative, which makes the offensive ones more noteworthy. In fact, many have been extremely enthusiastic and helpful.

As this is the third book project one has published on AGP, the takeaway here is that the AGP subculture of archaeology is demonstrably occupied by peculiarly turf-conscious inhabitants. Having established a niche for themselves, some AGP experts appear to be policing "their thing" against intrusions from outside interlopers. That is not a casual comment, and it is offered completely honestly and collegially to all readers, especially students. Mention that your study includes spatial data accumulated through the use of a total station and hardly anyone will raise an eyebrow; mention that you are working AGP into a larger anthropological research program and you might get some interested attention; mention that your explicitly anthropological focus is not AGP technology or theories of electromagnetic wave propagation or even how to spot underground appurtenances of aboveground megastructures and some seem to get angry or incredulous. Again, this is one's deduction, from nearly a decade of observation and experience. Something odd and patterned is at work in AGP publication and one is uncertain it is currently meeting the needs of AAs or Anthropology in general.

1.4 THE EVIDENCE AT HAND

A little more than 5 years ago, Conyers (2010) and Conyers and Leckebusch (2010) challenged geophysical archaeology to begin addressing cultural phenomena in AGP research and publication. This challenge has only partially been addressed in any manner, and many previous habits of AGP publication, especially the preoccupation with megasites and megastructures with surficial indices of buried materials, remain unaltered. Consider the quoted published abstracts listed below from some of the primary peer review journals, from during, prior to, and after 2010 in The Leading AGP Journal. These abstracts were chosen basically at random and feature several AGP techniques, and the European-style spelling was left in those cases in which it occurred. Notice the near absence of any focus on culture, behavior, or even people. Those absences are serious oversights.

> We report here a multimethod geophysical investigation of the Sant'Imbenia Roman villa archaeological site in northern Sardinia (Italy). The main objective of this study is optimizing a noninvasive approach to reconstruct rapidly the geometry of coastal sites. A hitherto unexplored area of approximately 700 m^2, adjacent to excavations, was investigated using ground-penetrating radar (GPR) and electrical resistivity tomography (ERT) surveys. The Sant'Imbenia villa is close to the present-day shoreline and subject to very high erosion rates and burial. A comparison of the high-resolution GPR and ERT models was made, and their integrated results are discussed in terms of providing a more complete picture that would not be attainable using a

single method. Geophysical analysis combined with archeological prospecting has revealed buried buildings north of the excavated part of the archaeological site. The results show that in this coastal environment ERT survey provided the most accurate reconstruction at the deeper wet levels of investigation.

Testone et al. (2015, p. 63).

Analysing the spatial distribution of anthropogenic relief structures can contribute to the understanding of past landuse systems. With automated mapping routines, small relief forms can be detected efficiently from high-resolution digital terrain models (DTMs). In this study, we describe an approach for the automated mapping of charcoal kiln sites from an airborne laser-scanning DTM. The study site is located north of Cottbus, Germany, where an exceptionally large historic charcoal production field has been documented in previous archaeological surveys. The goal of this study was to implement, evaluate and improve an automated GIS-based routine for mapping these features based on the template-matching principle. In addition to the DTM, different morphometric variables were evaluated for their suitability to detect kiln sites. The mapping results were validated against a comprehensive database of kiln sites recorded from archaeological excavations and via manual digitization. The effects of irregular kiln-site geometry and DTM noise were evaluated using synthetic DTMs. The results of the synthetic DTM mapping show that the template-matching results differed depending on the morphometric variable used for the mapping process. In accordance with this observation, a validation of the mapping procedure for the field site suggests that feature mapping can be improved. In particular, the number of false detections can be reduced using a combination of morphometric variables. For the validation area, the kiln sites with diameters of at least 10 m were mapped using the automated routine, with detection rates that were close to those of manual digitization. Therefore, the described method can considerably facilitate the mapping and distribution analysis of kiln sites or similar small relief forms that are prominent in a specific landscape.

Schneider et al. (2015, p. 45).

Geophysical data acquisitions in most archaeological campaigns aim to image the target structure directly. The presence of a target, however, may be inferred from its interaction with surrounding layers, if its relationship with those layers can be characterized sufficiently. In this paper, we show the use of ground-penetrating radar (GPR) to detect the subsurface continuation of the Ancient Egyptian tomb of the high-official Karakhamun (Theban Tomb 223) at the South Asasif tomb complex (Luxor, Egypt). Data were acquired using a Sensors & Software pulseEKKO PROsystem, equipped with antennas of 500 MHz centre-frequency, on a silty–sandy sediment surface directly over the target structure. A test vertical radar profile (VRP) suggested that the tomb superstructure was buried too deeply beneath sedimentary overburden to be imaged *directly: 500 MHz energy would propagate for only ~2 m before becoming undetectable. Attenuative layers within that overburden were strongly reflective, however, and could be used to provide indirect evidence of any underlying structure. When observed in the GPR grid, these layers showed a discrete zone of deflection, ~0.9 m in amplitude and ~4 m wide, aligned with the long-axis of the tomb. This deflection was attributed either to a collapsed vestibule beneath the survey site, or sediment settling within an unroofed staircase descending from ground- to tomb-floor-level; supporting evidence of this was obtained towards the end of the excavation campaign and in the following year. We highlight the value of such indirect imaging methods as a potential means of improving the capabilities of a given geophysical survey system, in this case allowing the GPR to characterize a target at greater depth than would typically be considered practical.*

Booth et al. (2015, p. 33).

The case study presented is a prime example of integrated geophysical— archaeological prospection. The aerial photographs available are complemented by non-destructive geomagnetic and geoelectric surveys with a reading distance of 0.5 m or less. To gain depth information and provide higher resolution, ground-penetrating radar (GPR) data are integrated. The GPR data were collected in a 0.5 × 0.05 m raster and visualized as black-and-white time or depth slices. The developments presented allow us to incorporate GPR into the standardized interpretation process of archaeological prospection based on a geographical information system (Grs). Using GPR and all the other prospection data available as a basis, a detailed three-dimensional interpretation model of the monument detected, the southern part of the forum of the civil town of Roman Carnuntum, is created.

Neubauer et al. (2002, p. 135).

The La Gila Encantada Site is located on an isolated ridge top north of Silver City, New Mexico in an area defined culturally as the Mimbres Mogollon region. The 180 m × 80 m (14400 m^2) sized site was originally recorded as a dense scatter of ceramics, lithics, and ground stone along with a number of surface depressions that appear to represent pithouses. Cesium magnetometer surveys were conducted to identify hearths, pithouse boundaries, and activity areas outside of pithouses in support of archaeological investigations, and to test this instrument's ability to image these features. This paper presents a characterization of the magnetic signature of a pithouse as a magnetic high in the center caused by the central hearth, low magnetic variability along the floor of the house, and then increased magnetic variability at the pithouse boundary. This characterization was successfully confirmed for three pithouses using cesium magnetometery and archaeological excavation.

Rogers et al. (2010, p. 1102).

This paper presents the results of fluxgate gradiometer survey of the Bronze Age city at the site of Kazane Hoyuk, southeastern Turkey. We undertook this

work to test the applicability of magnetometry to the study of the organiza-
tion of urban space at this site within the context of urbanization in Upper
Mesopotamia. Gradiometry collection covered a total of 37 520 m² in five
parts of the site. Results from each area were mixed but the most revealing
data, from Area 1, show a roughly 2 ha area in the outer town that contains
monumental, elite and administrative architecture as well as a main street.
Low negative values indicate that most identified architecture is built with
limestone foundations, and high positive values reveal that some of the
buildings burned before their collapse. These interpretations are supported by
excavations that reveal much about the use of the identified spaces and
features. Although the structure of Area 1 is rectilinear, evidence for strict rules
of city planning is lacking. Instead, the third millennium city at Kazane has a
structure seen at other Upper Mesopotamian cities: dense, semi-orthogonal
architecture built along well-maintained avenues. Combined with previous
research, it is clear that Kazane contained multiple elite or administrative
areas, which may indicate a degree of power-sharing or heterarchy in the
development and management of this city.

Creekmore (2010, p. 73).

1.5 SO WHAT *IS* AGP?

AGP should not by now need a finite definition, but here's one any-way: AGP is allegedly an Archaeological and/or Anthropological application of near-surface geophysics (Conyers, 2013). This book, however, is not really about geophysics as much as it is about using AGP to make observations of small-scale open-air sites, or h/g sites, most of which are truly old. As a workable definition for SSSs, one would suggest that these kinds of sites occupy indistinctly bounded areas in open, nonspeleological or structural contexts, and are composed of, or contain, archaeological materials and features of submeter-size to several meters in scale. For example, the French Paleolithic sites of Pincevent and Verberie could be classified neatly as SSSs by that definition. The Giza Plateau could not. Others may have different definitions, and that is to be welcomed and encouraged in order to broaden the discussion as widely as possible. One should add also that the artifacts, ecofacts, structures, and features of SSSs occur primarily in low-density (from solitary to several hundreds) accumula-tions of objects small enough to be held or manipulated by human hands. The preceding definition for SSSs is structured here so as to exclude the types of sites, structures, and materials that one would expect to observe on the Giza Plateau, or at Knossos, Harappa,

Poverty Point, Cahokia, Avebury, and similar monumental, mainly post-Neolithic sites. Such megasites are clearly not SSSs.

To this very day, the primary AGP literature contains a severe imbalance of focus upon the one type of site and not nearly enough attention on the other: in terms of Paleolithic SSSs the absence is nearly complete (as of late 2015: Conyers, 2013, personal communication). There is also a considerable bias toward the publication of European authors as well as biases toward publication of AGP applications at Greco-Roman, Iberian, Western Anatolian, and Egyptian megasites. We will discuss reasons for these imbalances and possible ways to rectify them, as well as focus on the abstracts from many publications to gauge other publication biases. Despite Conyers (2010) and Conyers and Leckebusch's (2010) calls for explicit AGP focus on establishing and posing anthropological questions to address, in effect constructing an "anthropological geophysics," it appears that the majority of scholarly attention remains firmly fixed upon AGP applications at megasites with large-scale surficial indices of buried materials. The preoccupation upon sites with such obvious large-scale surficial indices represents the *status quo ante* prior to 2010 even now in 2016. While Conyers (2010) and Conyers and Leckebusch (2010) sought to attract attention to new issues for AGP research, this call was apparently heeded to little avail as yet (Thompson, 2011, 2014, 2015).

CHAPTER 2

Recognizing and Defeating Biases

Why small-scale sites (SSSs) as an exclusive focus? Although the rest of the book will take up the challenge of elucidating that theme in depth, a brief introduction to some basic facts about archaeological geophysics (AGP) should suffice as a preliminary explanation. First, although AGP is not a "new" archaeological technique, its widespread application in the United States is new, at least in comparison to its use in Eurasia (Conyers, 2012, 2013; Thompson, 2011, 2014, 2015). Second, in response to a call for a more anthropologically oriented AGP from several of its more visible practitioners (Conyers, 2010; Conyers and Leckebusch, 2010), one has taken that particular challenge seriously and has attempted to assemble a coherent research strategy in response (Thompson, 2011, 2014, 2015), of which this book is a part. Lastly, some extreme publication biases are quite prevalent in AGP research and publishing, and these biases extend beyond particular scholars, publication venues, and continents (Thompson, 2014, 2015), which raises the suspicion that this heavily colors the reception AGP gets from "anthropological audiences."

What is an "anthropological audience"? By that phrase, one means merely a range of scholars who are attuned to the theory and method of Americanist Anthropology, as taught and practiced in four-field Anthropology departments as the holistic study of humankind. Not all archaeologists consider themselves to be anthropologists; most anthropologists would likewise not define themselves as archaeologists. Yet, in the United States anyway, it is extremely common for students of Anthropology to receive undergraduate and graduate instruction and training in all four of the following subdisciplines (in no particular order): Sociocultural Anthropology, Linguistic Anthropology, Archaeology, and Physical or Biological Anthropology. To be blunt, one can search many AGP journal articles and books in vain for any mention of Anthropology, and that is a problem. Specifically, the problem that one mentions relates to AGP's inclusion in US

Ephemeral Hunter-Gatherer Archaeological Sites. DOI: http://dx.doi.org/10.1016/B978-0-12-804442-1.00002-0

Anthropology and especially US archaeology departments, most of which assume that faculty and students both subscribe to an understanding of Anthropology as the holistic study of humans. While AGP can be used to address certain issues of archaeological significance, even a cursory review of AGP publications reveals that they are often only tangentially related at all to Anthropology and hardly ever premised upon the study of humans in any way (Thompson, 2015).

One has previously invoked the American rock and roll band Van Halen and their driving force Edward Van Halen as an analogy for use in exploring AGP reasoning (Thompson, 2015). There are a number of reasons for this. Anyone who has attended a Van Halen concert, or seen one on YouTube or elsewhere, will become rapidly aware of several things. One will note that whereas the music performed is most commonly presented as part of an ensemble, the featured instrument at a Van Halen performance is an electric guitar; the profound development and revolution of which has earned untold millions for luthiers, music companies, and guitar shops around the planet, and has been the subject of intense scrutiny from instrument manufacturers and guitarists for decades. Even the manners in which electric guitars are made, modified, processed, recorded, and used in performance (Thompson, 2015) have drastically changed. In the case of the band Van Halen and the musician Eddie Van Halen, you get the one only along with the other.

Had Eddie Van Halen not tortured and vivisected numerous poor guitar bodies, *and broke guitar builders' and players' rules to force outmoded old guitar manufacturing and playing styles into unprecedented new forms,* Van Halen would never have happened and the world would much be the poorer for it. In concert, the listener can hear Eddie Van Halen's highly derived, synapomorphic guitar used and intentionally misused and abused to produce sounds ranging from horses and elephants, to pianos and violins, to marching troops and many tortured noises that confound the definitions of music and that would frighten most banshees. Normally all that is involved are a guitar, a cord or wireless transmitter, and an amplifier. Things many people can get and use. And yet the removal of this instrument's wielder from a Van Halen concert would immediately invalidate the sonic experience for the majority of listeners. Very many guitarists can play Van Halen songs; only one untrained, unschooled guitarist on this planet can make Van Halen songs out of thin air and perform them to be what they are. Billions of dollars have been invested and recouped by guitar and amplifier manufacturers by analyzing or flat-out hiring Eddie Van Halen to assist in R&D. The why is simple: since the 1980s, millions of people ranging from teenagers to middle-aged guitar fans wanted to be at their homes and sound like Eddie Van Halen. The prices for genuine Eddie Van Halen guitars on online auction websites are absolutely staggering for what are merely pieces of wood, metal, and plastic (the priciest of which, at $25,000.00 per unit, are replicas of a beaten, worn Fender Stratocaster body cut into pieces and reassembled and with deep accumulations of ancient paint...very archaeological). What lessons can we draw from this?

First, we can learn that experimentation in any productive domain is absolutely necessary, including the intentional deviation from established themes, norms and conventions, and involving semi-adepts and accidental interlopers and intruders. Just because the in-crowd didn't invite you it doesn't mean that you shouldn't be at the table. Nobody knows from where the next great idea might emerge, or more importantly from whom. All are welcome and encouraged to tinker. Second, personal skill and some general knowledge are necessary but so are originality, enthusiasm, and creativity. Third, often the most crucial fundamental developments occur when "the rules" are held in contempt. "The rules," as it were, the clutching fingers of an Establishment, seek generally to preserve various prerogatives for themselves, usually small thrones situated at the nexus of money and influence; their directives should be heard and acknowledged, of course, but what they do and don't do, who they support and who they discourage, who they help and who they thwart, are often motivated by the most mortal of considerations. The next time someone discourages a good idea, remember Edide Van Halen: lots of people told him that he *had* to play in a certain way, he just *had* to use particular established guitar designs, he just *had* to record music in a certain way. There are always lots of *just have-tos* in Anthropology, Archaeology, and certainly we're told in publishing and living the academic life. In such ways are creativity and real innovation often stifled or completely suppressed, not always for legitimate reasons. Fourth, even the "best," tried and true designs and styles can be modified and improved if not completely innovated, and leaving such important tasks to the true believers and insiders are unlikely to result in any improvement or innovation. Anything can be advanced or improved where there is interest.

Conyers (2012, 2013) refers to improvements and advances in AGP in terms of computing power and miniaturization of computers and geophysical (GP) equipment, enhanced computer software, and various other important methodological and technological improvements. Some of these unquestionable refinements are concerned, however, with frankly better equipment than we should expect as manufacturers revise their products to keep buyers interested. Increasing methodological maturity in AGP is also something we should probably expect, as more archaeologists become more familiar with more AGP techniques and as these technologies become more firmly embedded in our general approaches to the human past. Obviously the physical and vocational/ practical "infrastructure" of AGP has greatly improved over time; it would be inconsiderate and wrong to claim otherwise. Yet, serious questions can be raised regarding whether most of the "easy" innovation and development has been materially accomplished to the detriment of advancing or innovating AGP *reasoning*. Better physical and methodological tools, and uses for them, don't of necessity indicate better theoretical perspectives. Has new thinking accompanied the evolution of new methods and gadgets? Have new authors been welcomed into the AGP publishing fold or are they scolded and criticized away?

Consider theory, in this case anthropological or Archaeological theory independent of electronic/magnetic, chemical, or physical theory used in traditional AGP publishing. To cite two recent "primary sources" from the AGP literature, Conyers (2013) and Johnson (2006), explicitly Anthropological or Archaeological theory is never actually addressed in these volumes. One has attempted to understand this phenomenon (absence of Anthropological theoretical content) over a span of years, through a doctoral dissertation and three published book-length treatments (including the present volume), and unfortunately it seems that further scrutiny results in less understanding of this matter. In seeking understanding, one notes that GP in general, regardless of the application venue (i.e., to geology, or soil science, or archaeology, or construction) conforms to various literary and sociological norms common to other remote sensing (RS) techniques (photogrammetry, LIDAR, aerial photography for several examples). It might come as a surprise to

Anthropologists and Archaeologists that RS and GP are used in all manner of environmental studies, urban planning, agriculture, construction, politics and economics, and many further uses. Although our present interest pertains to Anthropology and Archaeology, these are very limited and minority domains in relation to the total set of RS users. Many of those other RS users have developed entire literatures and publication norms all their own, some over much longer intervals than Archaeology, for instance. In terms of scholarly publication, those other literatures certainly exert a stylistic influence on the structure and content of AGP publications, because AGP publishing, at least in the United States, is still rather obscure and its writers needed analogs to serve as guides for how to develop AGP's literature body. Those guides came from other RS publications suited for often radically nonanthropological audiences. We can fix this and make AGP publishing more marketable for Anthropological users, although there are other complications.

Something that becomes obvious the more one immerses in AGP is that it isn't free and may not be the sort of thing many would like to see become common or inexpensive. This property may contribute to AGP's expense and to the peculiarly nonanthropological manner in which GAs publish, rendering their scholarly output difficult or inscrutable to read for most Anthropologists. One suspects that there are more than a few GAs who view AGP as a way to make money, including from their archaeological colleagues. It's hard to know how to evaluate this aspect of it. On the one hand, it is common for zooarchaeologists to hire themselves out to projects whose cohorts lack their faunal expertise. On the other hand, however, faunal analysis is a skill that can be acquired and developed far easier and for less money than the stupefying sums involved in acquiring geophysical equipment. If AGP became more widely and commonly applied in archaeology, and with a more digestible published corpus of literature, might that mean less money for the lucky few with access to the technology and with established reputations as AGP insiders? There are almost certainly humans in our archaeological midst who have entertained the costs and benefits of a widely accessible AGP and literature—and who have concluded that things are just fine as they are. One suspects this is also why it is so difficult for "outsiders" to break into AGP publishing as well. We need to humanize the process.

Why is an AGP literature thematically attuned to Anthropology and/or Archaeology even needed? Doesn't the one we already have work well enough as it is? No one can speak for all Anthropologists or Archaeologists, but in one's personal and professional opinion, a specific "anthropological" AGP literature is needed precisely because the one we have is not really "working" when the publications themselves are not useful in addressing humanistic research premises. For my doctoral dissertation, which involved an "anthropological" application of AGP in experiments and at a real archaeological site, there was no guiding literature to consult. We can improve AGP literature by providing such anthropological premises explicitly. We might ask ourselves just what humanistic avenues or areas exist for improvements in AGP thinking. This is asked intentionally to stimulate critical thinking and theoretical innovation. Innovation is certainly occurring *materially* in AGP, at least in applications at new sites and new kinds of sites; in applications involving new combinations of AGP technologies used at new and different sites; in combining AGP and GPS technology and data for geocoding and various other GIS applications. The assertion isn't that no innovation is occurring in AGP; what's asserted is that the previous focus has been to improve things other than the thinking of the people using the equipment and methods. Beyond advances in basic AGP technology, computers and software, "batching" or mating different techniques, and incorporating AGP with GPS and other spatial reference systems, what other improvements can be made and *where*? How much effort has been devoted to improving the *Anthropological* thinking of *people* using the technology?

How can we teach ourselves to ask better questions of AGP equipment and techniques? How can we teach ourselves to learn more about past human behaviors from AGP analyses of buried archaeological materials? Perhaps we should step outside Anthropology and Archaeology to inspect how GP is used in other activities. One particularly interesting use for AGP is in military applications: searching for unexploded ordnance (UXO) and mines. What do military personnel learn from using GP to locate UXO and mines? Although there are obvious hazard-mitigation uses through locating and disarming UXO, there is more than just what is found or not found, as we shall soon see. Once areas are cleared of UXO and/or mines, development and construction or modifications to those areas can commence. Humans and weapons systems and transports and other infrastructure can be employed in cleared areas. The military already knows all of this and much more, so I am not trying to help improve combat-effectiveness for anyone. There are undeniable proprietary interests in finding/disarming one's own UXO, for avoiding both civilian and military casualties and property damage, furthering maintenance of allegiance and protecting assets. We ought to ask, however, if this example actually constitutes "learning," at least of anything new. Military personnel generally have a rough idea which ordinance was used in what areas, so that might not actually be "learning," at least not at a high level. Military personnel can learn which proprietary ordnance failed to detonate and some of the circumstances of those failures.

In terms of enemy ordnance, similar lessons can be learned. In learning which proprietary ordinance failed to detonate, personnel can also learn possible causes for those failures, informing what not to use and where not to use it. Personnel can ascertain targeting or guidance and arming systems capabilities and weaknesses. Gauging failure rates and causes allows improvement of fabrication and other systems prior to deployment and live-fire activation. All of this amount to better intelligence in design, deployment, and uses, and fine-tuning weapons systems to particular situations and locations to maximize combat-effectiveness. Learning affects planning, and planning affects outcomes in this example. This is directly relevant to AGP in learning, planning, and driving outcomes, including further learning opportunities. Mines represent a similar GP application to UXO, especially in terms of detection and disarmament and avoidance or enemy ordnance, especially to reduce civilian impacts. However, analysis of mine location sites, depths, in what frequencies, and types deployed indicates a material index of human behavior, since deployment patterning reveals planning and intent. This knowledge indicates a great deal about enemy intentions and goals, speaking to the ability to learn something about how the other side thinks. Such thinking can also reveal goals, "values" (i.e., are civilian casualties and collateral damage planned and patterned or ancillary effects?) and other cognitive indices. Weapons deployments can affect human deployments and behaviors, with the goals of minimizing proprietary casualties and maximizing opponent damages. Knowing which areas contain UXO or mines, therefore, affects human behavior, planning, and contingency developments far beyond just knowing where and when to expect trouble and how to avoid it. Higher order thinking can be reconstructed from material indices as we see.

We can also potentially learn a good deal from the uses to which GP are put in the construction and demolition industries. In building, remodeling, demolition, and modifying roads, walls, floors, and other structures, it is common to encounter extreme hazards from utility lines, suspension cables, and other material dangers. Consider concrete sawing, cutting, and coring, for a particularized point of comparison. For those unaware of it, construction and modification activities often require the sawing or coring of concrete formations. These activities require the use of power tools that create high friction and sparks, employing carbon steel and diamond blades and bits to cut through concrete. Sparks and gas lines do not mix with safety, nor do cut live-wires or water lines. Cutting or coring through something like that can be expensive, destructive, or fatal at a job site. The construction industry has found it useful to deploy GP for utility location at job sites during development, modification, and demolition activities. Concrete-scanning for embedded hazards, in particular, is a particularly sophisticated application; Geophysical Survey Systems, Incorporated, e.g., has developed technology, methodology, and experimental testing and training programs devoted to concrete-scanning and utility location (the potential archaeological uses for this well-designed program are considerable).

Construction professionals have learned which situations can be dangerous and what can be done to mitigate hazards. This learning has resulted in better planning, improved construction and demolition techniques, and more personnel safety at job sites. Such learning processes work because intelligence, information, or "data," are applied a priori knowledgably to avoid pitfalls and failures. Consider also state DOTs and their use of GP in road and bridge inspection. For bridges and roads, GP can be applied to locate cracks, voids, and "rotten spots" (decomposed material from weathering and/or oxidation), areas of structural weakness or corruption. In conducting repairs and promoting general safety, such applications are extremely valuable. But much more can be and is learned. In analyzing data on construction techniques and material fatigue or failure, DOTs are able to learn which contractors build or repair things the best or worst; they can learn what techniques and structures are better-suited and where, for how long, under which conditions, etc. All of this only works if people teach themselves how to learn from the information that is provided. The equipment by which these data are provided are of drastically less importance than the data and possible conclusions themselves.

In regards to AGP, there are many relevant learning curves. Only the most superficial learning curve relates to particular individual mastery of GP equipment and ground conditions, interference sources, wave dynamics, and other considerations that relate basically to expertise in physics. We need also to consider the relevance and importance of the basic humanistic and paradigmatic, epistemological, or even theoretical assumptions we make Anthropologically, Archaeologically, and otherwise that affect our pursuits even before actual GP equipment is ever used. Although many of these themes have been mentioned previously (Conyers, 2012, 2013; Kvamme, 2003; Thompson et al., 2011), few published volumes have systematically developed Anthropological reference frames specifically designed to interpose AGP within the theoretical constructs we as Anthropologists and Archaeologists habitually employ. What is being sought? When is it being sought? How are we seeking? Why are we seeking? These and other important preliminary questions should not relate only or even primarily to GP equipment or archaeological materials.

Unquestionably, the future holds many unpredictable innovations in store for us, especially in terms of GP equipment sophistication, perhaps greater user-friendly interfacing (see the GSSI SIR-4000), and novel ensemble AGP multimethod applications. This wondrous potential future will also mean a profusion of AG publications similar to the ones we already have now: we tried it over here, it was "successful" look at the pictures, but what is the definition of success? This will not represent better thinking, however, nor will it be likely to facilitate a more widespread adoption of AGP in CRM or elsewhere.[1] We probably shouldn't expect a wider user-base for AGP among Anthropological scholars if what is learned either cannot be related to their own humanistic research or cannot be communicated to them. If innovation and improvement in AGP is confined to equipment and deployment techniques, it may not ever be directly applicable to larger Anthropological or Archaeological theoretical issues. It could be stuck

---

[1]This bears some special scrutiny and many people are interested in this but for different reasons. Numerous parties have advocated for a much better integration between AGP and CRM, the writer included (Thompson, 2014, 2015). One of the reasons for this is a potentially great utility and rapidity in conducting pedestrian surveys and documenting buried or unsampled sites (not always the same thing). It is also likely that AGP manufacturers and their marketing personnel view CRM as a growth market for plying their wares. Maybe some people also hope that CRM purchases would drive unit costs down, although this is unlikely because many AGP systems are handmade on an individual per-unit basis, not on assembly lines.

in a long and deep rut. Limited perspectives, fossilized thinking that perceives AGP as just a prospecting technique, just a RS method to identify major structural features, will produce limited outcomes. Maybe that's our biggest lesson so far: limited perspectives regarding the possible utilities to which AGP can be deployed result in the further production of limited output and even more limited perspectives, a circular reasoning structure as opposed to a dynamic field of inquiry.

CHAPTER 3

Seeing the Forest but Missing the Trees

Archaeological geophysics (AGP) is great for spotting roads, walls, rooms, houses, castle foundations, etc., at LSSs. AGP is also great for spotting Magdalenian lithic and faunal middens, probably hearths, and perhaps the remains of activity areas and domestic structures where present and in situ at open-air, ephemeral Upper Paleolithic sites (Thompson, 2011, 2014, 2015). There exists a great comparative collection of AGP imagery of buried roads and walls and other major rectilinear features to aid in the identification of similar features at other LSSs and nearly every new volume of The Leading AGP Journal adds more new examples of those same old uses. Is there a similar collection available for archaeologists who might like to use it to help identify buried lithic and faunal middens, hearths, and domestic structures at other small-scale sites (SSSs)? No, there isn't one and there are few publications even available on pre-Neolithic AGP. In terms of remote sensing, one could argue that we know more about the surface of Mars than about the AGP of open-air ephemeral SSSs on our own planet.

Perhaps the reader is doubtful. Consider this: How do astronomers and astrogeologists identify landforms in the geomorphology of Mars? Answer: (1) Observation and (2) Analogy. Such scientists use a variety of means to inspect data from Martian landforms, including radar and radio surveys. In fact, many of the same aeolian, fluvial, limnological, volcanic, erosional, and impact geomorphic features we take for granted on Earth have also been located on Mars, including rills and gullies, meandering and braided stream channels, basins and lakebeds, dunes, dust devil tracks, stepped shorelines, volcanoes, and impact craters. They have been identified through remote sensing techniques similar to near-surface AGP (Comins and Kaufmann, 2005). How are such distant features *identified*? Most people with functioning visual ability could see interesting things on the Martian surface. But what process allows such things to be identified with high confidence through

Ephemeral Hunter-Gatherer Archaeological Sites. DOI: http://dx.doi.org/10.1016/B978-0-12-804442-1.00003-2

remote sensing? Analogies are what provide us with confidence about such identifications, in this case analogies with virtually identical terrestrial landforms. We know how impact craters and fluvial channels are formed on Earth and we use analogical reasoning to infer that they were formed by the same processes on Mars. If scientists can use radar and radio data collected from another planet to identify landforms with high accuracy, why can't we use similar technology to identify material signatures of human behavior at SSSs on our own?

What is largely lacking from traditional AGP publishing and research? Discussions of ancient people are lacking, for one. Logical Anthropology rigor is lacking; the systematic structured argumentation in the manner of Social Science, pertaining specifically to Anthropological Archaeology reasoning (Gibbon, 2014; Thompson, 2011, 2014, 2015). Notwithstanding rigorous discussions of electromagnetism, geophysical equipment and theory, and physics, there is little rigorous discussion of actual people and Anthropology in most AGP literature.

There is little to no anthropological or even archaeological epistemology involved in most traditional AGP publication. Published AGP research is instead focused mainly on geophysical theory and not anthropological theory. Traditional AGP publications seem preoccupied with "seeing" rather than "seeing as" or perceiving (Gibbon, 2014). This relates to a very big difference between traditional and AGP archaeology in terms of basic observation. "Seeing" is simply a matter of perception, while "seeing as" requires much more refined training and specialized knowledge (Gibbon, 2014). Seeing evidence of buried archaeological material at any scale in AGP imagery is a matter of pattern recognition and equipment experience (Conyers, 2013; Thompson, 2011, 2014, 2015). Recognizing the possible behavioral nature and disposition of buried archaeological material should be a more formal and informed process of perceiving the signatures of past human behavior in that patterning—"seeing as." At present, using traditional AGP methods, "seeing" is possible with actual explanation requiring ground-verification—the expensive and arduous task of excavation. We should focus on developing theory and methods for approaching subsurface archaeological deposits in a process of "seeing as"; understanding observations as indices of human behaviors that we can learn to recognize through MRT, the use of comparative image collections, excavation experiences and maps, the anthropological archaeology literature, and other references. Simply locating buried material is neither Anthropology nor Archaeology.

For example, many archaeological sites are located during construction activities. Does that make construction an informed, appurtenant subdivision of Archaeology? This is not to make light of AGP or prospecting, but rather just to note that the development of ever more refined means of detecting buried artifacts and publishing accounts of them are not, strictly speaking, Archaeology. It is also certainly not Anthropology even when many of the elite AGP practitioners occupy Anthropology departments (at least in the United States). Contemporary AGP publications appear to assume that AGP represents a stylistic variant of Archaeology, similar to zooarchaeology or ethnoarchaeology, and yet one can search in vain for *any* mention of culture, behavior, or even humans. It is almost as if finding and describing the *appearance* of buried things, minutely explaining the settings and parameters in software used to create images, and calibration of equipment used to gather data for the making of pictures, are AGP's premises.

AGP *could* be used in more sophisticated manners than merely thing-finding, providing certain material patterns have been linked to certain human behaviors. Similar patterning at the same or even other SSS can be provisionally inferred to be similar in behavioral cause; different material patterns could relate to different contingencies. Material signatures could be defined as the middle-range theory (MRT) linkage between observed material conditioning and inferred or demonstrated human behavior. Perhaps excavation would belie such inferences in some cases, but for those that are not falsified, the linkage would soundly remain. At any rate, such behavioral inferences are far more robust than the typical AGP conclusions that "Something is here" or "Something is not here," or "Conditions are not conducive for AGP application." The strength of the material signature, or its coherence, depends upon the strength of the association between material pattern and behavior; the frequency of pattern occurrences and their visibility; and pattern repetition across space at the same site, or at different sites. Such inferences and linkages could only be made much more robust by a strong experimental replication program, published AGP literature relating to it, focusing on imaging SSS and characteristic material phenomena at them. By this method, we can (1) reduce our ignorance about comparative AGP material visibility, configuration, scale, shape, size, frequency, distribution in SSS material phenomena; (2) reduce our ignorance in distinguishing between actual archaeological material and false-positives in AGP data; (3) identify anthropogenic intrusive material disturbances.

Although some AGP authors claim that such methods are uniquely useful in "landscape archaeology," using AGP simply to fill-in a given landscape with prospected, located sites for later excavation or cataloging fails to address other data that could potentially be extracted. Merely noting where things are or aren't does not really address many behavioral correlates with observed or imaged material phenomena. While knowing where things are or aren't is useful at a cursory level of analysis, a presence/absence survey is anthropologically and humanistically uninformative even when plotted spatially or geocoded or entered and manipulated in GIS applications. The use of advanced equipment doesn't inform us about past human behavior unless we focus on the behavior through the use of the equipment.

CHAPTER 4

Material Signature of Human Behavior: Framing a Hunter-Gatherer AGP Agenda

What do humans do to leave detectable material traces behind, whether over the short-term or longer durations? The reason we concern ourselves with material evidence is because the archaeological geophysics (AGP) technologies we discuss in the book detect material phenomena. Because that's what the technologies "see," we adjust ourselves to what they show us. While the phrase "material signatures (MSs) of human behavior" might sound imprecise, it is actually a good descriptor for what remote sensing does and what we do with it. For example, for decades many different countries have employed aerial reconnaissance, from balloons and dirigibles to fixed and rotary wing aircraft, to missiles, remote control drones, even orbiting satellites, as a means to detect enemy troop and military material movements about landscapes. People move things, these movements are related to other tangible activities: they leave materially visible traces on the ground, or in vegetation, and can be used to project future personnel or ordnance positions, as well as gauge future behaviors and even retroject past ones. We can use MSs to know what people are doing and in many cases what they did in the past.

The only thing that is really "new" about the use of MSs in the ways proposed in this book relates to their application specifically at small scales and at ephemeral archaeological sites. For all of the reasons already covered previously, this represents a very understudied aspect of AGP research, one perhaps at variance to many of the assumptions we make in relation to uses for AGP in the field. In some cases, we should also specify that we may not be able to "see" direct evidence of human behavior in the archaeological record via AGP. What we might see are various patterns and associations and patterned absences that we have, elsewhere in our research, learned to correlate with certain past human behaviors. Thus we are seeking

Ephemeral Hunter-Gatherer Archaeological Sites. DOI: http://dx.doi.org/10.1016/B978-0-12-804442-1.00004-4

both direct material evidence of past human behavior and certain material phenomena we can correlate with others through AGP applications.

I have elsewhere (Thompson, 2014, 2015) described the process outlined above as a form of middle-range research (MRR), or middle-range theory (MRT), in which explicit linkages are made between archaeological material phenomena and past human behaviors that could have impacted them. Consider the sorts of materials and their spatial arrangements and patterns we can expect to see at small-scale site (SSS): possibly small, submeter-sized hearths or lenses of charcoal and ash, various configurations of animal bones, lithics, and fire-cracked rocks distributed in space, voids or empty spaces between such configurations lacking artifacts, hut impressions, or compacted subresidential "floors," etc. All of these are potentially visible to multiple AGP methods at many ephemeral sites. Yet, until we know what such patterns "mean" at the level of individual sites, ground-verify the actual material identity of AGP phenomena, and link them to actual human behaviors, we haven't done much besides prospect, or look for buried things.

How can we get the most out of a search for MSs without also needing excavations? This is a tough question to answer, unless we realize and admit to ourselves that all AGP applications, unless explicitly ground-verified, are simply digital approximations of reality, digital models, and possible representations of buried material phenomena. Unless we intend to ground-verify, or expose, each and every phenomenon revealed by AGP technology then we have to admit that they are just models, perhaps very good models, but models nonetheless. So the question we should be asking is: How reliable can we make our models and our abilities to evaluate them? We have to assume that, unless we misuse or damage it, our AGP equipment is functioning and revealing to us basically accurate models of subsurface phenomena. We can also train ourselves in the interpretation of wide ranges of AGP data from numerous different contexts. We can train with verified experts or manufacturer-sanctioned trainers in the use of such technology. Yet, none of those actually address the models themselves or what we can do better to learn from them and inform ourselves.

Although it isn't normally made explicit, one of the primary research designs in AGP is the use of inductive reasoning, as we have described it, to go from specific to more generalized contexts or levels of analysis. Inductively, we would seek to learn as much as possible from a specific MS at a particular site so as to extrapolate knowledge gained to other sites, or to a general corpus of knowledge, applying it into an established methodology. An excellent example of conducting archaeological (in this specific instance, ethnoarchaeological) research inductively is lithic replication studies. In such research, primarily self-taught flintknappers manufacture, or "replicate," a given type of stone tool, presumably by using materials and techniques that were actually used by the Ancients. The replicated item is then used experimentally to perform a range of duties, and its relative effectiveness in performing such tasks is recorded in exhaustive detail. This implement might then be subjected to microwear studies or other technical analyses to ascertain wear patterns unique to certain tasks, document breakage or attrition, and other mechanical or chemical examples of use-wear. Many such studies have very profitably resulted in the acquisition of much hard data related to uses and effectiveness of many very ancient stone tools.

While we cannot (or at least should not!) use AGP equipment in such ways, there are available means to perform an incalculable range of purely experimental studies of material configurations characteristic of SSS and to evaluate the models AGP portrays of them such that we can develop better MRR and MRT and refine our cognitive approach to AGP data. We can extrapolate what we have learned at one site and use it in relation to AGP data from others, at least generally, to give us access to ranges of past human behaviors that have been correlated with certain nonphysical behavioral agencies (Thompson, 2014, 2015). We could also experimentally make things AGP has never seen at such sites, and gauge its usefulness in imaging them. Besides, what we are after is not so much imagery of particular material as it is the behavioral inferences we can make from them. It's the ranges of possible behaviors that we are really trying to isolate and not merely just patterns in data.

We have numerous examples, from Binford's (1978) research on Nunamiut ethnoarchaeology to Kelly's (1995) more comprehensive study of foraging societies, of correlates, of linkages, between human behavior and material phenomena. Such basic data can be experimentally compiled in any number of ways, at various depths, scales, in unique sedimentary and soils contexts, to which AGP can be applied productively to inform us about many of the things we don't yet know, and may never know beyond experimental means. From the sources mentioned previously, and more, we have many examples of excavated archaeological materials in firm spatial associations and integrity, to inform us about how material at SSS tends to be patterned. We also know quite a bit about patterned absences of archaeological material at sites, and have cogent explanations thereof. All of this and more can be easily exposed to AGP surveys in experimental "sites" so we can inform ourselves about wider ranges of geophysically resolvable archaeological manifestations at SSS. This could even be an exciting, inclusive open-source research program available to all archaeologists, or at least those interested in AGP. But none of it will happen or reach its potential without active participation from students and instructors and even CRM professionals in an organized program.

4.1 HOW DO WE SEE MSs?

As described in other chapters, "seeing" and "seeing as" are two very different propositions and processes (Gibbon, 2014). How do we proceed from seeing material to seeing MSs of human behavior? It isn't primarily an MRR or MRT cognitive process, although such steps are necessary to link our material observations to presumptive past behaviors. Rather the first requirement is simply a willingness to use AGP as a platform for making observations about MSs of past human behavior. We would be looking to develop means of using geophysics not as a mere thing-detecting apparatus but as a platform for performing "ethnoarchaeological" observations of the archaeological record, which means changing the way we approach the equipment and changing how we think about AGP. A willingness to see AGP in such a way and to participate are perhaps the most important steps to take.

Obviously changing attitudes will also be an important part of such a program. A stale fixation on AGP as merely a thing-finding or thing-sizing apparatus is unlikely to advance such pursuits. Nor will active or even passive resistance from the AGP Establishment help. One suspects that an ethnography of AGP practitioners would be a very valuable addition to the growing AGP literature in order to expose, even anonymously, many of the assumptions, goals, and self-interests of that community. Likewise, an anthropological study of AAs who have consulted/are consulting with GAs would be unquestionably informative and would provide solid data beyond merely anecdotal offerings. Such research would lay bare many of the attitudes we might seek to alter regarding AGP in a changing Archaeological and Anthropological world, to say nothing of a planet and a human species in constant flux.

Conyers (2012, 2013) has suggested that changing attitudes among archaeologists and geophysicists are in fact necessary. Such revisions are necessary to open minds to new possibilities, as well as to open fields of inquiry to new practitioners, to new scholars with new ideas. So we might "see" this volume as yet another attempt to invite a wide cohort of interested parties into a new and collegial program to provide AGP with an Anthropological basis, a humanistic as well as physical premise, so that anyone who is interested in using it can do so productively, not merely prospectors and technicians. It would be immeasurably helpful if even more AGP types were engaged in the project. Precisely that reason is why I have not tried to be overly comprehensive in outlining this program of "observational AGP." This could be a collaborative and cooperative effort, or at least it ought to be.

And yet, at any rate, "seeing" MSs *as* MSs will require different metrics, different methods at different sites and different times. There won't be a "one-size-fits-all" method that works everywhere, and hopefully not just one or even a privileged few will be able to dominate it. In fact, because the archaeological material inventory of SSS can be so highly variable, the more people working on this the better, the more eyes with which to see. But the focus on SSS is actually necessary, not merely because of their general absence in the scholarly AGP literature but more practically because material behavioral signatures are more clearly visible in isolation with them. It's hard enough to disentangle MSs at LSS even through painstaking excavation, let alone to see them via AGP. SSS can actually feature unique MSs that are not conflated or compressed and altered into indistinct wholes.

Mainly, all this program requires is for people to adjust their thinking slightly about the ways in which we can use AGP at some sites (not all, it is important to note). This relates to the "seeing" versus "seeing as" dichotomy. To some extent, many of us cling to the view that AGP is simply a prospecting device as opposed to an actual observational platform. It can actually be both, and this "taxonomy" or classification for how AGP relates to the larger disciplines also extends to how we "see" AGP. Perhaps we should work to change or sophisticate the archaeological taxonomy into which we place AGP. Consider how we can get trapped by our own taxonomic constructs, by our own heuristic devices. We are seeing just this in relation to much of the Linnaean classification scheme.

Without getting bogged down in comparative specifics, we can perhaps learn some lessons by considering the ways in which our improved technical expertise with DNA has undermined the Linnaean classification system. That scheme has worked quite well and for a long time. Yet, with regard to many animals and their actual genetic relationships, we can spot cracks in the Linnaean edifice: recently grizzly bears, Kodiak bears, and polar bears were found to be interfertile and capable of breeding and producing reproductively viable offspring, thereby making them, according to the Modern Synthesis and the definitions most people use, different geographic populations of the same species (Mallet, 2008). As can be shown in relation to the presence of Neanderthal DNA in the modern human genome (Sankararaman et al., 2014; Vernot and Akey, 2014), it begins to seem as though what we call a "species" is a much more fluid biological entity than Linnaean taxonomy can accommodate, and this pertains to human origins research as well. As can be seen in any number of published and even anecdotal and media-driven scrambles for alternative hypotheses, some researchers have returned to using concepts such as "subspecies" and some also to dangerously and politically unwise reiterations of "race" as post hoc accommodations to new genetic frontiers. While providing opportunities for new knowledge, this also illustrates how we can become bound to earlier, simpler concepts in the classification of things. One hopes we might see a similar diversification in "seeing" AGP "as" more than thing-spotting technology for showing large structures; we can also see it as an observational platform from which to observe variable MSs of past human behavior in some ephemeral settings too.

In many ways, it seems ironic to issue calls for cooperation and scholarly engagement relative to a subject that is experiencing intense scrutiny. AGP is becoming more integrated in more American archaeological projects than ever previously. More archaeologists are at least passively familiar with the series of techniques involved and the possibilities they offer than in earlier intervals. In Europe, such techniques represent a significant proportion of archaeological projects and play a role even in many projects in which they are not the primary rationale. Yet, many dedicated archaeology types, many excellent scholars, are frustrated, impatient, even discouraged in regards to AGP, and this can certainly be puzzling but also quite motivating. To quote something I have learned to find inspirational:

> Again, you can't connect the dots looking forward; you can only connect them looking backwards. So you have to trust that the dots will somehow connect in your future. You have to trust in something—your gut, destiny, life, karma, whatever. This approach has never let me down, and it has made all the difference in my life... Your time is limited, so don't waste it living someone else's life. Don't be trapped by dogma—which is living with the results of other people's thinking. Don't let the noise of others' opinions drown out your own inner voice. And most important, have the courage to follow your heart and intuition. They somehow already know what you truly want to become. Everything else is secondary.
> **Jobs (2005, http://news.stanford.edu/news/2005/june15/jobs-061505.html).**

The quote above is from Steve Jobs, shortly after he was diagnosed with terminal pancreatic cancer.

AGP is unlikely ever to become as ubiquitous and fixed to contemporary life as the iPhone and its numerous derivatives. But we can find cause to pause, and to reflect, by reading and *seeing* Jobs' words above *as* an invitation to press boundaries, to push envelopes, to agitate against stale orthodoxies. It is exactly the asking of questions, often uncomfortable ones, that drives our Sciences and our species. If we don't ask questions, if we simply accept orthodoxies and the status quo, among other things we are guilty of failing to take important matters seriously. If something is a serious matter we have a duty, an *obligation*, to question it mercilessly, lest it become just another dogma.

Since this is partially a pupil-centric volume, it is important to highlight an example of the power of unstable orthodoxies and why endless scrutiny of orthodox positions are necessary to our discipline (Thompson, 2012). It is an honor to teach. Not only is it not a dishonor to teach, it is also not some subordinate prerogative relative to what many term "pure research" in most arrogant treatment in the academic world. Research is a personal indulgence supported by excellent timing or excellent ability in convincing people to spend money; teaching is a calling. A lot of Anthropological currency and epistemology and "pure research" has recently died on the altar of genetics in the past 5 years. For several decades, although the position had been adopted for many decades before, the orthodox Anthropological explanation of "modern human origins" proceeded from the authoritative pronouncement that Neanderthals and "anatomically modern humans" were different hominin species. These species could not interbreed because they were different species. Even lacking any substantive genetic evidence of such an orthodoxy, the orthodox archaeological position was that the material evidence "supported" the preconception that Neanderthals and modern humans were separate species and could not interbreed.

For students who are entering Anthropology now, this is a wild and great time. Because we now know that Neanderthals and modern humans did in fact interbreed, multiple times, in multiple places, and we know this because we have numerous human samples that show clearly that Neanderthal DNA is found in the contemporary non-African genome (Sankararaman et al., 2014; Vernot and Akey, 2014). Students will get a lot of double-talk and a lot of post hoc accommodative reasoning to explain this in relation to the bare bones of their evolutionary teaching. But the facts are the facts: as we now understand "species," as populations of closely related interbreeding organisms, humans and Neanderthals are the same species. It is difficult to argue Neanderthal found in the genomes of non-African people without admitting that the archaic DNA got there by a most human method. It is also difficult to suggest that these taxa were separate species unless one invokes special pleading to explain why humans are "different" genetically. Those are the old rules, and by the old rules if Neanderthal DNA is found in modern human populations then they are the same species.

The coming years will witness many peer-reviewed and peer-approved attempts to reconcile the recent genetic discoveries with the backlog of material archaeological inventories. These will include attempts to explain how Neanderthals were culturally stunted even if they were genetically compatible with "humans." Such things are already entering the professional literature (i.e., the idea that we weren't separate species, but that we just hadn't had enough time to become complete reproductive isolates...changing the subject to avoid the question) and knock-off media (NOVA), etc. See, humans live in a cultural-ecological niche; our environment is basically our culture. In that medium, the construction of little subniches situated at the nexuses of influence, notoriety, money, and tenure become very valuable, and things that threaten those linkages are to be avoided. Real manual labor and poverty is often considered the result of a failure to police those valuable apices. Manual labor, poverty, and loss of "prestige" are things to avoid studiously, which means uncomfortable questions are discouraged.

Science has never been an occupation for the faint of heart or the passive. The giants of our way of thinking often risked their very lives to do their research and disseminate it to other people; they usually confronted enormous Establishments (and still do) premised on exactly the contrary to their research conclusions. One can attain a lucrative position moving papers from one side of their desk to another, pushing buttons and pulling levers, assiduously not asking questions easier than one can find a spot in Science. This is the way of our unfortunate world. But instead of dissuasion, for those intrepid and courageous of heart this is an invitation, a request, a plea, for the curious mind to become involved in Science. If scientists tell you your idea is bad it probably isn't; it's just something they didn't think of before, something they cannot privatize to themselves, something they cannot control. And that includes the scientists employed by governments to police their fields for grants and stature. Carl Sagan (2005) explained how and why they herd us in such ways. It's not really for the benefit of Science or even of scientists. We should try to mitigate that, in whatever ways we may.

Science does not belong to scientists, nor does it belong to anyone in. Science belongs to the planet, to the world, to all of humanity, or it belongs to no one. This is a dispersed model of ownership, one much more attuned to the post-Internet world of open-sourcing and collaborative ventures, crowd-funding, and mutual sharing in general. Unquestionably, therefore, some people will loathe it and that description/definition of it, because if something can't be monopolized, oligopolized, and it can't be easily modulated to channel resources in certain directions, it seems "uninteresting" to them. We should be aware that although this mecentric model of magnified self-importance exists in many places in US society, it is not the only model, and it certainly isn't conducive to increasingly diverse academic disciplines, including Anthropology. Carl Sagan (1995) wrote eloquently about how poorly Science can be mismanaged for personal acquisitiveness and financial gain. One often wonders how Sagan would perceive contemporary American academia in the 20 years since he died.

Analyzing the AGP literature, one can become overwhelmed by the alleged "complexity" of data interpretation. One ubiquitous overriding theme seems to be, "Interpreting AGP data is vexingly complicated and ambiguous, so we should probably leave it all up to the experts." This is convenient. It reinforces the mistaken impression that geophysical theoretical experts, and not Anthropologists, are the unquestionable authorities on AGP. Yet, AGP as it is currently practiced is not strictly archaeological and certainly not explicitly Anthropological, especially in theoretical terms. It's more like "geophysical sociology," where the aftereffects of highly organized state societies are the majority of its subjects of inquiry. It speaks little if at all to the long interval of human hunter-gatherer (h/g) lifeways on the planet, which span of millions of years as opposed to several thousand in the case of urban AGP. Worse, the sporadic and infrequent literary presence of Paleolithic AGP simply confirms the current confirmation bias that big stone monuments are better (easier).

Here I return, yet again, to the need for a systematic and organized program of experimental simulation AGP research especially devoted to material phenomena characteristic of small-scale ephemeral sites. The market is saturated already with trials of AGP gear on megastructures, to the detriment of exploring the majority of small, ephemeral h/g sites accumulated during the longest interval of our evolution on this planet.

I advocate ephemeral simulations in order to calibrate *us*, not tweak equipment settings to make prettier pictures in a journal. If geophysics can be used to locate concrete interior hazards—which are some specific instances of dangerous/destructive hidden potentials or MSs of past human behavior (which they are)—then it can also be used for other contingent, presence/absence and associational, morphological, scalar, configurational, etc. comparative investigations of past human behaviors. Allow me to return again to GSSI and their training program for concrete scanning. GSSI trains many nonspecialists to use GPR in extremely sophisticated ways to produce useable, *valuable* data, and information for avoiding concrete hazards for construction and demolition activities. More is the point: this training does not require explicit theoretical backgrounds in physics, geophysics, or even any science in particular. Personnel lacking any specialized scientific training can productively use such equipment very accurately.

Consider the following. Suppose that we can learn how to locate and *identify*, via AGP, Paleolithic stone hearths at *suitable* (where ground conditions allow it) open-air sites through the use of simulation experimental AGP training and our knowledge of the published real-world examples from the archaeological literature. Notice I did not write *all* Paleolithic hearths. Perhaps a reliable remote method for locating discrete ash lenses from simple ground fires will never develop: they don't really tell us much anyway except that someone once set a fire. But stone hearths do; they indicate an investment in a space, humans contributing to making it a place. Let's stipulate that we can teach ourselves to locate and identify Paleolithic stone hearths at an 80% rate of positive location and identification after ground-verification. So what? What do stone hearths tell us? Enter Anthropology. Suppose we identify and ground-verify a Paleolithic stone hearth somewhere. Building a stone hearth takes time, effort, and energy even if all the rocks and wood (or other combustibles) necessary are nearby and easily accessed. What do they tell us about the people who made them? What if one was alone? Did solitary hunter-gatherers bother to build reusable stone hearths? We've already started building inferences. Where are the rock outcrops that sourced the rocks in the hearth (we might wish to find out after we ground-verify)? How far away from the site are they? What is the intervening topography, what terrain was traversed with the rocks in transport? Now we're looking at hints of labor organization and plausible numbers of individuals involved, and even in a cursory examination these are solid, important archaeological data. Concrete data, as it were. Data not obliterated by hundreds or thousands of years of continual human activity obscuring the signal as they are at megasites with monumental structures.

Since Paleolithic stone hearths were made by humans, if we could somehow analyze every single Paleolithic stone hearth ever made and quantify our data, we would probably find that their measurable attributes could be plotted graphically as a normal distribution with strong central tendency. Given what we know therefore about people and stone hearths, we can learn how to ID them reliably in AGP data. Suggestions that we ignore scale, configuration, shape, and other material characteristics in AGP data (Johnson and Haley, 2006) are utterly puzzling since those are exactly the sorts of things concrete scanners use to spot interior hazards. If we throw out the baby bath along with the water and the baby what have we got left to analyze? MS doesn't refer purely or simply just to arrangements of material and speculation about what caused them. MS also involves the Anthropological theoretical and epistemic premises or larger archaeological research agendas, analytical techniques, and the inferences used in reaching conclusions.

We can also similarly devote ourselves to locating and identifying artifact scatters and "middens." We can combine accurate, to the items or features, simulations of such phenomena and compare them to actually published archaeological data with known artifact numbers and deposit thicknesses, with materials ranging from bones, to lithics, to FCRs and thus use AGP the same way. Such middens can provide anthropological data as long as we take the time and make the effort to teach ourselves how to correlate AGP data and images with material phenomena and behavior. Humans tend toward symmetrical regularity, which is a huge advantage for us. Regularities in shape, size, material, configuration, and other criteria abound. The fact that GSSI, for just the one example, can train people reliably to identify interior concrete hazards—in order to prevent cutting things like gas lines, live electrical conduits and other things that shouldn't be cut— ought to be a humbling datum for the aware. Minute theoretical awareness of physics and wave dynamics is not necessary—the equipment takes care of the physics while people find and identify formal features and phenomena present in the data.

CHAPTER 5

Human Material Signatures

Although this is a book on Archaeology, in many ways we don't have to look only at the remote human past to find relevant examples of material signatures that we can use to refine our views and ways of seeing human impacts on material *as* indices of human behavioral phenomena. While we will discuss geophysics in terms of the platforms from which we might wish to make such observations about past human behavior, we should consider just how common such material indices are in our own contemporary society, often not restricted even to human events. In the Midwest United States, it is very easy to tell when thunderstorms have occurred: tree branches, entire parts of trees, lots of foliage, and occasionally many dead birds killed by falling hailstones will be seen; tree removal services will be in abundance, with wood-chippers mulching tree remnants. Iowa cornfields provide visceral and stark evidence of the passage of tornadoes, even funnel clouds that skim over the fields without touching down: sinuous tracks of twisted and flattened stalks evince the proximate passing of spiraling vortices. Human examples? Old trees—mainly stands of deciduous species, but old conifers as well—planted in rectilinear shapes around old Iowa farmsteads reveal the past preference for the planting of windbreaks and ornamentation around such properties, even when the residential and agricultural structures have long since been demolished; pioneer cemeteries, with eroded limestone headstones, often few in number, associated churches no longer in evidence, stand as stark reminders of very different Midwestern settlement patterns in the past. Such signatures abound if we know where or—perhaps more importantly, how—to look for them. Just seeing them isn't enough: it is necessary to look deeper, *to see them as* material signatures of past human behaviors (Gibbon, 2014).

Consider the author's very small late-1800s hometown of Newburg, Iowa. Today, one can walk around this unincorporated "village" and find abundant evidence that it was once a bustling hub of human residential and economic activity. Although the village consists now of only five actual streets and a steadily decreasing number of humans

Ephemeral Hunter-Gatherer Archaeological Sites. DOI: http://dx.doi.org/10.1016/B978-0-12-804442-1.00005-6

and their residences, numerous concrete foundations surficially visible indicate the former presence of railroad yards, a grain elevator and cooperative, a lumberyard, a creamery, a post office, a grocery store, a bank (complete with the preserved foundation of the actual safe clearly visible in the middle of the structure), and a blacksmith shop (on the author's own property). The smithy, in particular, although completely demolished decades ago (including the foundation) still continues to produce numerous surficial artifacts; nails, hinges, horseshoes, iron rods, twisted lumps of metal, bricks, chunks of concrete, iron slag, coal cinders, and all manner of broken glass panes and containers strongly indicate its former location. An especially quaint reminder of this once active locus of human behavior is the presence of former public sidewalks preserved in some areas, poured long ago with mixtures no longer used for such purposes today. The driving force that underwrote all of those long-gone human activity loci, the railroad, still runs right where it has for perhaps 150 years, occasionally even now bearing locomotives pulling grain cars and corn oil/corn syrup, but no longer raw materials such as lumber, iron, etc. Newburg's decline as a small but active Midwestern town can probably relate directly to changes in the economic base; as fewer trains stopped in Newburg, instead heading to the much larger railroad hub of Marshalltown, Iowa about 25 miles to the northwest, where materials could be offloaded and redistributed in much larger bulks, Newburg began to die. Those economic changes led to later changes in human behavior, settlement pattern, and even residential relocation.

Such phenomena are repeated wherever humans lived in the past and live today. Our activities frequently leave traces behind us consisting of modified or displaced material and associations unique to the activities we perform. Even at the small-scale of hunter-gatherer family bands and individuals, so long as such material expressions are not destroyed by natural physical or chemical processes—and especially not overwritten by later expressions of more intensive human activities, we can expect to find many of them. How can we locate them? To locate such material expressions, we have to train ourselves to identify the frequently symmetrical patterns in which they can be found. This is called "pattern identification", and it is the basic method and process whereby fingerprinting, satellite reconnaissance, medical tomography, driving, and many other frequent human activities are possible.

5.1 ARCHAEOLOGICAL GEOPHYSICS FOR HUNTER-GATHERER ARCHAEOLOGY

Diversity is supposed to be a valuable Anthropological concept and a unique perspective of our discipline. We have seen, however, that there isn't much diversity at all in the subjects of most archaeological geophysics (AGP) publications. Hunter-gatherer archaeologists of the Paleolithic and Mesolithic and those working in Paleoindian and Archaic periods, or even specialists on the few remaining foraging cultures of today, could be forgiven if they have the impression that AGP just "doesn't work" for their brands of Archaeology. Yet, we wouldn't have to wonder where such a mistaken impression originated when we reflect on the overt biases present within the primary AGP literature. Hunter-gatherer archaeology matters, and it matters a great deal at that; but not, apparently, in AGP. For about 99.99% of the time humans have existed, we lived foraging lifestyles, making, using, and living in a wide continuum of small-scale sites. Humans are today who and what we are as a function of evolving inside a pervasive foraging adaptation. Over an interval of more than 3 million years, most archaeological sites ever formed were foraging sites of one type or another, some not yet even formally described, many more not even located (or, worse, won't ever be located because they are on continental shelves and were submerged after the ablation of the Pleistocene ice sheets and subsequent sea-level rise). So it is very odd that scant focus and attention is devoted to hunter-gatherer archaeology in the AGP literature.

Is no one trying to do hunter-gatherer AGP? Of course they are! Most of the submissions probably just don't make it past peer review. To a degree not seen in most other Archaeology publishing, names and places matter in AGP, and the same names and the same places recur over and over again in that literature. And the leading authors in AGP are not interested in foraging archaeology. This has been made very clear. Why? One suspects it has to do with the interests of most AGP reviewers and their ignorance of, e.g., Paleolithic Archaeology, which doesn't feature large building foundations and walls or castles or other big things that are easy to see with human eyes. Paleolithic archaeology consists structurally of cave sites, rock shelters, and open-air sites, with simple ash lenses from ground-fires, stone hearths of different types, artifact "middens" and other accumulations, pits, and small features and phenomena, often invisible at the ground surface. Perhaps such things do not excite megastructural AGP experts; maybe stone hearths that formerly held precious, warming fires for their sparse, huddled makers leave the AGP experts of today feeling cold. But to a hunter-gatherer specialist, a hearth is not always just a hearth. Consider stone hearths, or slab-lined hearths to name just a few types. How much about people can we learn from hearths once we teach ourselves really to see what we are seeing?

For the pudding is not always in the proof when foraging site manuscripts are submitted and denied publication in AGP venues. I have participated the past 4 years in the writing of three major, formal peer-reviewed manuscript submissions based upon real Paleolithic AGP applications, with abundant data, strong imagery, clear disciplinary currency, and obvious literary import. Two of these manuscripts were submitted to the leading AGP journal, the other to the flagship American Archaeology journal. All three of these papers were premised upon the establishment and refinement of methodology for identifying actual archaeological materials at Paleolithic sites using AGP technology and providing useful methods also for eliminating spurious nonarchaeological or nonanthropogenic materials present in the data therefrom. I should also add: in these papers we explicitly referenced our goal of making AGP user-friendly for other Paleolithic archaeologists, so as to prevent them from needing to reinvent wheels we'd already invented, just in case they might want to employ AGP on their own in their own ways. All three of these papers were reviewed caustically, unprofessionally, and, much worse from an holistic disciplinary Anthropological perspective, ignorantly. The reviewers admitted that they had no idea what the point of such hunter-gatherer material was. This came as a shock until a deeper understanding was obtained. The point was not that we had written badly, but that *the reviewers could not understand why establishing valid methodology for identifying stone hearths and middens at Paleolithic hunter-gatherer sites and how to tell real archaeology from mere rocks and other materials in the data through simulation training and live-fire field application was important.* The reviewers not only didn't understand what they were readingbut also didn't care.

For myself, I am interested in Roman aqueducts and triumphal arches, but I don't find them personally inspiring. I also have some problems with a breathless and undue haste to investigate monumental architecture to the practical exclusion of mere Paleolithic or other hunter-gatherer sites in AGP publications. If humans hadn't survived their Pleistocene trials, would there even be any Roman monuments or villas to survey via AGP? The fact that humans are alive at all today is something that inspires me, and we only accomplished that by passing some pretty severe Paleolithic thresholds. So methods for identifying hearths in AGP data—and seeing them as hearths—are actually tremendously important, all the more because they are often all we see at Paleolithic sites! We can learn much about human biocultural evolution from such apparently uninteresting, humble human structures: stone hearths imply reuse and more than just a few hours' or an overnight's stay. Sourcing the stones informs us how humans were moving within landscapes, how many people were involved in their formation, and many other useful Anthropological data. The puzzling AGP focus on what is essentially the Historical Archaeology of cities is, by comparison to the mysterious and elusive Paleolithic, utterly droll given the presence of voluminous historical records and sources that often describe the commission and building of the things themselves.

Please peruse the AGP literature. Please. I think endless and repetitive articles about what are basically huge elite structures like castles are absurdly boring when we can just observe how human elites operate today, without heading to the field or republishing the same basic manuscripts continually. Perhaps this literary bias—it's actually much more than just a simple observation bias, it's more like a preoccupation or a unique form of colossomania—is due to Geophysical Archaeologists (GAs) being much more Geophysicists than Archaeologists. But should people who are primarily geophysicists—lacking exposure and training in small-scale Paleolithic Archaeology, often intimately linking humans and their environments—review and decide which aspects of Archaeology are important or publishable? I don't think so.

Perhaps I labor in vain. Perhaps I have located a solution for a non-existent problem. Maybe the European bias toward AGP as being nothing besides a sounding method for delimiting the extents of Roman villas and the ruined foundations of Byzantine churches in Nabataean cliff-cities is simply too established, just too great. Inertia, momentum, and all that. The nearly built-in historic archaeological bias in AGP publishing suggests quite brusquely that Paleoindian, Mesolithic, and any other hunter-gatherer archaeologists who might want to use AGP for purposes that suit them need to find their own publication venues, dedicated to what interests us. The post-Neolithic urban AGP folks, those Eurasian city-entranced-people, are evidently uninterested in what we do with our rocks and our bones.

Perceiving what is observed as opposed to just observing is the basic dynamic I am describing. Considering the Paleolithic situation I describe above, it appears that there are entire varieties of Archaeology, whole intervals of time, for which AGP cannot be much more than a side-interest for the Archaeologists working inside them, at least not until AGP and its established authorities adapt themselves and their literature and *perspectives* to those of Anthropology, Archaeology and humanism (see also Thompson, 2015). I am describing a focus on people and their behavior, not gadgets. What most AGP research and publications feature today represent "comparative near-surface geophysics," with Archaeology serving merely as a stage upon which the actors perform.

What are the end-goals of prospecting or AGP surveys? Are they just to locate excavation foci and to eliminate presumed unproductive areas of the subsurface? If just locating things to dig is the main purpose of AGP, then that also implies that anything potentially archaeological located in such a survey has to be excavated before we can say anything about it, and this is highly irresponsible on the one hand and wrong on the other. We have to force AGP beyond its assumption that absolutely everything has to be ground-verified, because that means, by logical extension, excavating any archaeological material located by it before saying anything conclusive about the AGP data drawn from them.

Should we not also teach ourselves how to make the best inferences possible from potential archaeological material located in AGP surveys so as not to excavate everything? What do we owe Archaeologists of the future, who might be armed with better ideas and methods than we have now? If we ground-verify everything, what will be left for them?

CHAPTER 6

Pattern Recognition and Preliminary Identification

At small-scale site (SSS) or in locales where multiple SSS have been established to feature repeated instances of obvious and similar material patterning in scale, configuration, and shape, excellent opportunities could exist to refine techniques for preliminary pattern *identification*. Although Johnson and Haley (2006, p. 306) expresses doubt that ephemeral site features (such as trash pits) like shape or scale will be useful in the identification of archaeological material in AGP images, these properties are actually extremely important for a program of pattern identification as well as recognition. Our abilities both to recognize and identify patterns in buried archaeological material are skills that can be developed. In a general sense, over time, much refinement in these pattern recognition (PR) and PI skills will result cumulatively simply as more archaeologists acquire the necessary experience. Specifically, however, this process involves inductive reasoning—moving from specific instances to more generalized contexts with ephemeral SSS features. This inductive endeavor will require us to determine what is representative of given classes of ephemeral features (pits, hearths, middens) not just in graphics from published literature but also from experimental trial and error, simulations, and other activities. We should never assume we have "seen" examples of all hearth or midden forms. This is where we need broader simulation training with experimental reconstructions or replications of SSS archaeological features.

To fill in gaps in our knowledge, experimental simulations will be necessary. There should be many opportunities for students and really anyone interested because one's suspicion is that purely experimental middle-range research for AGP is regarded as low-status by the "high-profile" personalities. We should actually be encouraged to do this. The nuts and bolts of Science are not always events of Einstein/Hawking notoriety, and there is a serious need for dedicated systematic simulation research. One of the world leaders in the manufacture of ground-penetrating radar (GPR), Geophysical Survey Systems, Inc., uses many

Ephemeral Hunter-Gatherer Archaeological Sites. DOI: http://dx.doi.org/10.1016/B978-0-12-804442-1.00006-8

concrete simulations to train personnel to locate interior hazards. Furthermore, GSSI also trains their own employees using such simulations. If GSSI does it, why aren't more archaeologists? Using simulations of known materials surveyed by blind students would also be a great way to develop PR and identification skills.

Of course, the development of such a program assumes that numerous people participate in it, which is not now the case. Who has the dedication to Archaeology to contribute to such a general training program? The desire for profit and prestige and tenure thwart this process of disciplinary development, which ironically might thwart a more widespread acceptance or enthusiasm for AGP. Participants will need a willingness to conduct important research that nonetheless may not be publishable in prestigious archaeology journals.

The simulation experiment program could also lead to the development of "trait-centered archaeology" explicitly for AGP (Gibbon, 2014). The goal of such a process would be the development of familiarity among interested parties with the "traits" (shapes, sizes, configuration, distribution) of the archaeological record characteristic of SSS. Much actual data from such sites still needs collection in addition to experimental simulation. This is not descriptive of taxonomy or cultural-historical archaeology, nor is it "social" or postmodernist archaeology searching for "meaning." At its most basic, it is induction, learning from specific instances with understanding applied more generally. What "traits" or features (or absences) are common to SSS? What is unique to them? What material patterns are more or less shared globally?

Much contemporary AGP publishing features a large measure of intentional over-technicality, focused on pure physics and many other things unrelated to Anthropology. If we are not supposed to look at the shapes and scales of subsurface material in AGP data (Johnson and Haley, 2006) then precisely at what are we supposed to look? Flint, e.g., a natural form of siliceous glass, has dielectric properties similar to those of siliceous (quartz) sands. So if flint lithic scatters deposited on quartz sands tend to reflect AGP waves in characteristic ways, illustrative of similar shapes and scales of material deposits, when they reflect patterned signals, should we really ignore apparent patterning they produce in AGP data? For even in the event that AGP data indicates nonanthropogenic material conditioning—or later instances of intrusive anthropogenic disturbance of archaeological deposits—do we really just ignore what our senses and brains tell us?

Our senses *work*, and they are really good at recognizing patterns in raw data, especially data patterned through the prior activities of other people. Our brains are really good at extracting patterned information from the data they receive from our sensory apparatus. Evolution has trained our physical brains, the genetic code that builds them, and the activities by which they learn over our lives to do exactly this—to recognize patterns. So we really can't just discount patterned subsurface material information we observe in AGP data—especially when that's exactly what we seek to see in such data!

If lithic implements—including simulated lithic implements made by latter-day flintkappers—that show patterning in their shapes, scales, and reduction sequences were regarded by the same observational paradigm that governs AGP data, we might have to conclude that all lithic taxonomy is subjective and unreliable. If we didn't see a particular Acheulean handaxe made in real-time, we don't and can't know anything about it. If Paleolithic zooarchaeology reveals identifiable patterning in faunal remains at archaeological sites, then according to AGP norms we'd have to conclude that faunal analysis is subjective and unreliable. This is an example of the exclusive and defensive AGP community behaving stubbornly. On the one hand, AGP folks say AGP is useful for archaeology; on the other hand, according to the "experts," AGP data is so fraught with uncertainty that its interpretation is utterly subjective and unreliable. It cannot be both ways. While AGP data can be open to variable interpretations, it does not mean that people cannot learn how to interpret it. The argument that AGP data is subjective and unreliable is actually a tacit invitation for more people to become involved in interpreting it!

I am writing to reach out to people who want to use this powerful technology in ways that interest us in Archaeology. Hunter-gatherer (h/g) Archaeology—especially high-resolution h/g Archaeology—is where many archaeological techniques were first developed—especially forensics and CSI methods. If the Archaeology of small-scales is missing the point, beside the point, or is generally uninteresting to the AGP community, then presumably so are the later methods to which it gave birth, including those used to solve crimes. I am reminded of a very interesting forensic AGP article a dear AGP colleague sent me a few years ago. Solla et al. (2012), which is in many ways an exemplary article on AGP forensics methods, experimental simulation, and for interpreting GPR wave dynamics, and understanding how the radar waveform changes as it passes through different materials, illustrate a common point of divergence between the Gas and the Anthropological Archaeologists (AAs). How many dirt Archaeologists understand how radar waves (microwave radiation) attenuate (speed up and slow down) depending upon the dielectric properties of the materials through which they pass? Wave dynamics are interesting to physicists, but the implications for small-scale AGP applications are probably what would interest most h/g Archaeologists, and it is the physics, and not graphics indicating how their simulated subsurface targets appeared in the data, that Solla et al. (2012) stressed. AAs would probably care more about productively and accurately interpreting plan and profile reflection models generated from the data than waveforms, and this makes sense because they are concerned more about where things are and aren't than they are about how materials differentially attenuate microwaves underground.

CHAPTER 7

Simulation Sandbox

For part of my doctoral dissertation, using an example from Leckebusch and Peikert (2001), I built a 2.5-m^2 sandbox out of ¾ in plywood from a local lumber yard. The depth of the sandbox was exactly 1 m. Starting from the bottom of the empty sandbox, I filled it to a depth of 10 cm with pea-sized riverine gravel from a nearby construction aggregate company (Martin Marietta, Colfax, Iowa). Atop the gravel, I built a simplified version of a Paleolithic stone hearth out of four cobbles of oolitic limestone from a now abandoned, nearby stone quarry (Ferguson, Iowa). The hearth was a symmetrical 40 cm^2 in diameter. In depth, it was approximately 15 cm thick. Over the gravel substratum and the simulated hearth, I placed homogenous fine quartz sand 90 cm in depth, bringing the total sandbox sediment depth to 1 m. Although I placed a number of smaller and less massive simulated archaeological features in the sand at varying depths, imaging the hearth with archaeological geophysics (AGP) equipment was the primary goal, in order to see what it would look like. The reasoning was simply that because a stone hearth is probably the most diagnostic recognizable feature at smallish Paleolithic open-air sites.

My colleagues and I used The University of Iowa's GSSI SIR (subsurface-interface radar) 3000 ground-penetrating radar with both the GSSI 400-MHz and 900-MHz antennae. The photo of the hearth in Fig. 7.1 shows as it appeared in the visual spectrum, while the two ground-penetrating radar (GPR) time-slices, one at 11 ns time-depth and the other at 10.5 ns time-depth, show high-amplitude radar reflections of it. The images I include were produced by the GSSI SIR-3000 and the 900-MHz antenna. I forego any discussion of the SIR-3000 settings because they are superfluous to this project and to this discussion. What matters is the GPR's ability to resolve and image my simulated hearth, and using the 900-MHz antenna it performed admirably.

Although I could have played with the GPR images some more to make them "pop" better in more contrastive color palettes, the included

Ephemeral Hunter-Gatherer Archaeological Sites. DOI: http://dx.doi.org/10.1016/B978-0-12-804442-1.00007-X

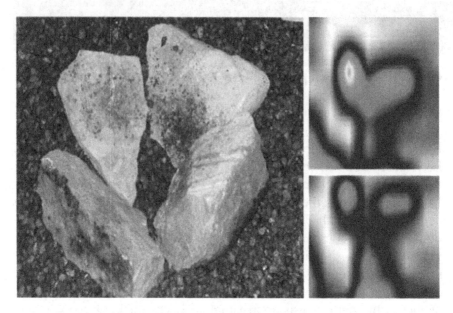

Figure 7.1 Simulated Paleolithic Hearth 40 cm² in diameter. GPR time-slices at 10.5 ns and 11 ns in time-depth at right.

images clearly show the little stone hearth in my sandbox. This is not a matter of subjective interpretation, because the simulated hearth was the only pseudofeature I constructed in the southwest quadrant of the 2.5-m² sandbox at that depth. Some other pertinent particulars: my sandbox was built inside my garage, an old carriage house built along with my house in 1901. The sandbox was built atop a 6-in concrete slab with no reinforcing rods or other metal inclusions, just a plain slab of concrete. Because the sand was dumped outside my garage in my driveway and was rained on several times before I put it in the sandbox, it retained a fair amount of interstitial moisture at the time it was surveyed with GPR. There was nothing else above the sandbox, outside it, or below it that could have interfered with the radar or produced the images of the hearth except the simulated hearth itself. These data were collected October 2007 and May 2008.

This is the sort of project to which I refer as simulation middle-range theory (MRT) experimental AGP research in this book. The entire cost of the project, including plywood, some 2 × 4-in pine studs for framing support for the sandbox edges, corners, and sides, sheetrock screws, and the sand and gravel (about 7.5 tons worth of sediment) cost about $350.00. And despite what Johnson and Haley (2006) suggest, the shapes and scales of the simulated archaeological material in my sandbox were what I expected and extremely important to me: the high-amplitude reflections produced by each of the buried "targets" I put in the sandbox looked very much like the actual shapes and scales of the materials just as I "deposited" them in this exercise. Since all buried "targets" (I use the terminology targets because I was shooting radar waves at them, hoping to hit them and to bounce reflections off of them intentionally) except the hearth were extremely thin (only a few centimeter in thickness), they only showed through several time-slices taken from the "surface" of my simulated sediment column down to the concrete floor of my garage. As I progressed in time-depth down through the slices, familiar shapes would suddenly appear for one or two time-slices of 1 and 0.5 ns in time-depth, and then just as suddenly disappear as the depth in focus progressed below them. If the reader wishes to see more of this material and the extensive AGP data presented in my dissertation, you can find it on Pro-Quest online through a college or university library access point. Unfortunately, these are not free to the public. The main upshot here is that shape and scale of subsurface materials matter *profoundly*.

7.1 OTHER EXAMPLES OF RELEVANT SIMULATION TRAINING

In August 2015, I was fortunate enough to take a GPR course at GSSI, Inc. in Nashua, New Hampshire, taught by Mr. Dan Welch. The course was entitled, "Theory and Practice of Applying Subsurface Interface Radar in Engineering and Geophysical Investigations." In this course, the students were given GSSI SIR-3000 to use along with the 400-MHz and 1.6-GHz "blue cart" antennae, and the fantastic all-in-one GPR and antenna ensemble unit, the StructureScan Mini HD (2.6-GHz antenna). We surveyed, or "prospected," outside on the GSSI lawn, out in the neighboring streets and parking lots, and inside the GSSO corporate facility on a garage-type deep concrete slab and in an appurtenant room, using various concrete sample slabs of differing thicknesses, dielectrics, all with very different material inclusions inside them. We also surveyed the cinder-block walls of the GSSI building, and were able to "see" the voids inside the blocks, some with re-bar inside them other lacking rebar. All of the different materials produce somewhat or very different reflections in the radar data. We were taught how to use 3D data collection techniques to produce accurate 3D images using GSSI's proprietary RADAN 7 software. I also collected some data in San Diego during further concrete-scanning training and collected more data from a friend's home garage concrete floor slab in Danville, Indiana (Thanks John!).

The images I collected in San Diego consist of an underground storage tank (UST) for caching diesel fuel at the Penhall Company San Diego branch, covered with an 8-in concrete slab with heavy rebar and various pipes and conduits one would expect at a fueling station or other similar location (Figs. 7.2−7.4). I collected the San Diego data along North−South and West−East transects. Since I knew I was dealing with very large metallic features, I did not bother with a 3D collection method. One can clearly see in radar profile the top of the N−S profile, as well as the approximate actual shape and dimensions of the UST and the rebar in the W−E profile. The reader can also see with the W−E profile that I played around with the horizontal and vertical scales of the radar data using GPR Viewer Plus, which I got from Dr. Larry Conyers. Exaggerating the vertical scale and squashing the horizontal scale enhances the imagery to differing degrees, the active process being to play around to see what works and what looks best you to.

For the residential garage floor slab (Fig. 7.5), I used the GSSI StructureScan Mini HD on the slab in my friend's garage in Danville, Indiana. I got the idea for this by looking at some of GSSI's own concrete-scanning training literature, where they claimed that it was possible to distinguish between "fresh" and "rusty" reinforcing rods

Figure 7.2 GPR image of wire mesh in concrete training floor at GSSI, Inc. Nashua, New Hampshire. Imaged area is 2 m².

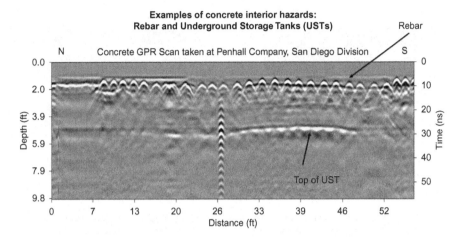

Figure 7.3 GPR North-South profile of an underground diesel fuel storage tank.

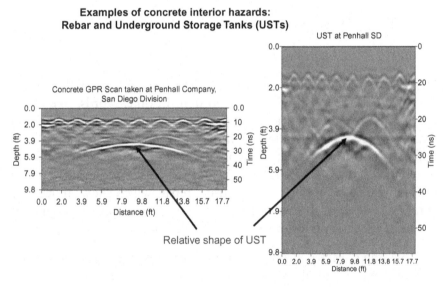

Figure 7.4 West-East GPR profile of an underground storage tank. Vertical axis exaggerated in image at right.

inside concrete slabs. The image I produced from data I collected at GSSI's concrete-scanning facility floor clearly shows a neatly delineated wire mesh mat with distinct edges to the wire. The image I collected from my friend's concrete garage floor in Indiana, most likely poured by a previous owner in a do-it-yourself type project, not only showed the image I include with the "fuzzy" rebar wire mesh mat edges, indicative of rusty rebar according to GSSI's own training

Other examples of materials and conditions in concrete

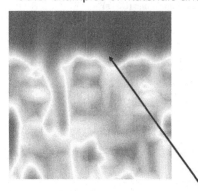

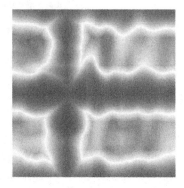

Above is an example of two superimposed slabs in a residential garage of concrete poured at different times with different reflectivities. The dark gray mass is the boundary between layers.

Above is rusty rebar inside the same slab at left, in a garage probably set and poured by owner.

Figure 7.5 GPR time-slices of 7 and 8 ns time-depth in a garage in Danville, Indiana. Area shown 1 m².

literature, I was also be able to see where one portion of the garage floor slab onlapped above another portion. In other words, one of the previous owners of my friend's house had poured his garage floor in multiple stages, and the interface between the concrete poured at those different times was visible; probably because a film of water had infiltrated the contact between the two different parts of the slab. For me the really interesting thing is the "rusty" rebar, which shows reflections perhaps related to the oxidation of the raw steel wire mesh and its chemical effects on the surrounding concrete matrix.

Now, the dielectric properties of raw steel rebar and flint lithics are very different, as are the scales, configurations, shapes, sizes, and manner of accumulation in subsurface environments. But some of what the above (and much more) has taught me is that often very subtle features and items are visible to GPR, and therefore probably to other AGP techniques as well, quite distinct from their typical applications at huge sites on enormous materials and structures. And since this is not a book about wave dynamics, differential dielectrics, remnant magnetism and the like, I am not including such information. That would be like Eddie Van Halen telling us exactly how he set each parameter of his guitar rig when he recorded certain songs, equalization, etc., not to mention postrecording signal-processing EQs and other effects and techniques in the recording process. Signal-processing—during and after both recording music and AGP data collection—can drastically alter sound and AGP data. I like to try and "find" Eddie's sounds on my own with my gear, and in the same vein I think it's important for hunter-gatherer, small-scale archaeologists to experiment with AGP and discover what works best for them.

CHAPTER 8

Material Signature of Seasonality at Verberie

The graphics referenced are available online at the following URL: Fig. 8.1 is a 20-m^2 reindeer bone density plot in Golden Software Surfer 9.0 that shows bone frequency per unit area from Secteur 190 at Verberie; no lithics or Fire-cracked rock (FCRs) are plotted. The circular bone deposit is clearly visible. One must remember that along with the bones were excavated thousands of lithics and numerous FCRs from the same midden, or, perhaps more likely, disposal pit. This feature was excavated in 1976, and was attacked as a single deposit. Elevation data for the individual artifacts were not recorded so precise provenience is not recoverable. This is what archaeologists mean when we say that we kill our informants by excavating, because once done it cannot be undone, so it's best to record all pertinent data during excavation. Yet, the scale (approximately 2 m^2) of this feature, its configuration, its shape, and its basic demeanor stand in stark contrast to the similar Fig. 8.2, also a 20-m^2 reindeer bone density plot, which shows the bone frequency per unit area in a midden from another "sector," Secteur 202 Level II.22, about 10 m northeast east of the VBC 190 plot (again only bones were included in the plot). The scale and configuration of the bone deposit in Fig. 8.2, as well as the general semicircular shape of the midden, are rather at variance to Fig. 8.1. Although the scale, configuration, and demeanor of these features are obviously quite different (one is a dense circular pit, the other a laterally diffused semicircular midden) what does that tell us?

There are two primary points illustrative of the material signature value in these figures: (1) they are qualitatively and quantitatively distinct in terms of much more than just their relative shapes (more on that below) and (2) Fig. 8.2 betrays a very close similarity to actual AGP data collected in 2006 prior to the excavation of the fauna portrayed in 2009. Fig. 8.2 portrays a semicircular material mass that produced high-amplitude reflections in the 2006 ground-penetrating radar (GPR) survey at VBC. It is likely that the feature in Fig. 8.1 would also have been sensibly captured had it been subjected to AGP

Ephemeral Hunter-Gatherer Archaeological Sites. DOI: http://dx.doi.org/10.1016/B978-0-12-804442-1.00008-1

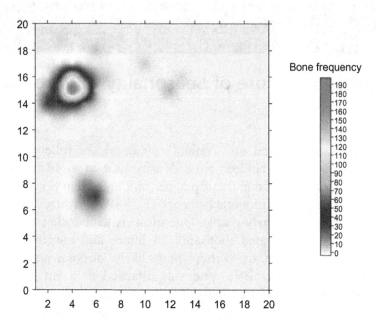

Figure 8.1 VBC Secteur 190 reindeer bones, teeth, and bone fragments plotted in Surfer 9.0 per unit area (frequency of fauna per 0.5 m². Scale is 20 m², tick marks at 2-m intervals. Note highly circular symmetry of this deposit.

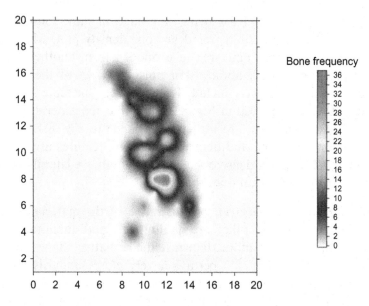

Figure 8.2 VBC Secteur 202 II.22 reindeer bones, teeth, and bone fragments plotted in Surfer 9.0 per unit area (frequency of fauna per 0.5 m²). Scale is 20 m², tick marks at 2-m intervals. Note semicircular configuration of this deposit and its resemblance to Fig. 8.3. This is actually fauna that were recovered from VBC using the AGP data in Fig. 8.3 as a guide. It was only later that this patterning was "seen as" (Gibbon, 2014) a material signature of differential human behavior.

Figure 8.3 12-m² ground-penetrating radar time-slice at 25 ns time-depth from 2006 VBC field season, using GSSI SIR 3000 and 900-MHz antenna; X- and Y-profiles at 12.5 cm intervals. This is the same area shown in the bone frequency plot in Fig. 8.2, prior to excavation during the 2009 VBC field season. Note the semicircular material configuration above and its similarity to Fig. 8.2. This was used as a guide for where to excavate during 2009, and the high-amplitude reflections (bright colors (gray in print version)) above corresponded to recovered fauna, lithics, and FCRs.

survey prior to its excavation; recall that the feature in Fig. 8.1 was composed of numerous lithics and FCRs not present in the bone plot. What this means is that, using AGP at VBC, we can "see" differential material signatures in the spatial, scalar, and configurational attributes of artifacts relative to the human behavioral conditioning that caused them. Even assuming no ground-verification commenced from such an exercise, one could still infer that such recognizably different material patterns visible in AGP data were indicative of even deeper potential differences in the material forming them.

Since the archaeological deposits in Figs. 8.1 and 8.2 were excavated, we know what they are and more importantly how structurally and functionally distinct they actually were in real human behavioral terms. Figs. 8.4 and 8.5 below portray interesting differences in the faunal age class profiles from the two faunal assemblages plotted in Figs. 8.1 and 8.2. The VBC 190 and 202 II.22 reindeer teeth were subjected to a dental wear analysis and the differences were surprising. About half the caribou from VBC 190 were not just senior citizens when they were killed; they were absolutely ancient in reindeer years. In contrast to this, the VBC 202 II.22 caribou population was younger,

1 m

Figure 8.4 1976 photograph of the circular deposit from VBC 190 during excavation; note the articulated reindeer cervical vertebrae at center-right, virtually an entire caribou neck preserved nearly as-was (from Thompson, J., 2011. Views to the Past: Faunal and Geophysical Analysis of the Open-Air Upper Paleolithic Site of Verberie. Unpublished Doctoral Dissertation on file at The University of Iowa and on ProQuest and UMI., p. 140, Figure 4.6). This would've been radar-visible had AGP been applied to it prior to its excavation as judging from 2006 surveys and extensive archaeological material recovered in 2009 from VBC 202 II.22.

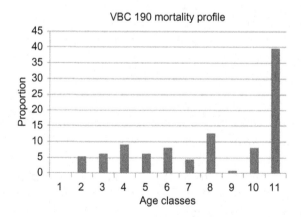

Figure 8.5 Dental wear age classes from VBC 190; note the very high skew toward geriatric caribou. Nearly half the caribou from VBC 190 were very aged animals. From Thompson, J., 2011. Views to the Past: Faunal and Geophysical Analysis of the Open-Air Upper Paleolithic Site of Verberie. Unpublished Doctoral Dissertation on file at The University of Iowa and on ProQuest and UMI., p. 175, Figure 5.19.

with more prime-age animals present and a much lower proportion of geriatrics. We should also note that, according to Audouze (2016, personal communication), the VBC 190 fauna is chronologically later than that from VBC 202 II.22, and were positioned several centimeters higher stratigraphically, so it should be inferred that these two faunal assemblages were also temporally distinct and not associated by shared human occupations. So we have some clues: the VBC 190 caribou were very advanced in their lives when they were hunted. These geriatric reindeer were dispatched, butchered, and disposed of in such a manner that they were piled in a circular heap or pit. The caribou from VBC 202 II.22 were much younger and were killed, butchered, and disposed of in a laterally dispersed semicircular midden. Why would that be the case? (Figs. 8.6 and 8.7).

At this point, we have to consider the differential nutritional states of these two very different reindeer populations. As one who has "scavenged" (for data, not for food!) over 100 white tail deer carcasses in Iowa for my own research over the past 6 years, I can state quite emphatically that there are profound differences in the quantity and quality of the bone marrow from older and younger white tail deer, and by inference caribous as well. The frequently harsh Iowa winters not only cause the deaths of many deer but also preserve their carcasses for the intrepid (if somewhat morbid) observer. Cracking open the medullary cavities of young white tail deer appendicular long bones normally reveals abundant creamy whitish pink or slightly yellow

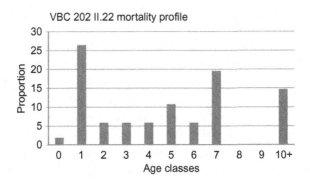

Figure 8.6 Dental wear age classes from VBC 202 II.22; note trimodal distribution of age classes and the higher proportion of younger caribou in contrast to Fig. 8.4. This was a much more youthful faunal population at time of death. From Thompson, J., 2011. Views to the Past: Faunal and Geophysical Analysis of the Open-Air Upper Paleolithic Site of Verberie. Unpublished Doctoral Dissertation on file at The University of Iowa and on ProQuest and UMI., p. 177, Figure 5.21.

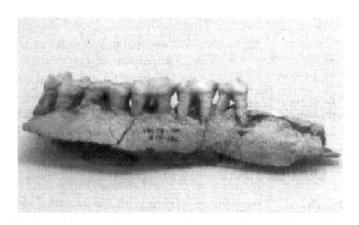

Figure 8.7 Example of one caribou left mandible from VBC 190 with articulated 4P-3M. Note extreme molar wear, hence advanced age of this individual at time of death. This specimen is on display at the Antoine Vivenel Museum, Compiegne, France. From Thompson, J., 2011. Views to the Past: Faunal and Geophysical Analysis of the Open-Air Upper Paleolithic Site of Verberie. Unpublished Doctoral Dissertation on file at The University of Iowa and on ProQuest and UMI., p. 143, Figure 4.9.

marrow deposits with a rubbery texture, high in fat and oil content. The youthful, pliable marrow completely fills the medullary cavities in the respective long bones. Repeating this process on geriatric deer produces pale white or even a sickly gray marrow, with a spongy, grainy, or brittle texture, indicative of lower nutrient, but especially lower fat and oil content. The aged marrow only partially fills the medullary cavities. The marrow from senior white tail deer also often simply falls apart in one's hands. Even when their deaths may have been caused by nutritional stress, young deer bones produce bone marrow that is higher in quality and quantity than bones from geriatric deer.

Another key difference between the caribou from VBC 190 and VBC 202 II.22 relates to the human treatment of the bones themselves. Although the two datasets were similar in terms of frequency (1610 bones, teeth, and fragments for VBC 202 II.22 to 1855 of the same for VBC 190), the proportions of identifiable specimens are roughly 5:1, skewed heavily toward VBC 190. So not just the spatial disposition of these faunal assemblages was different, the fragmentation of the bones themselves due to human activity also differed substantially. Enloe (2004, 2006) and Thompson (2011) indicate moreover that there are few indices of mechanical bone sorting or destruction present at VBC, so something other than mere decomposition and taphonomy explains the different treatments to which these fauna were subjected. We can also infer a cause for this discrepancy. The highly fragmented VBC 202 II.22 caribou bones, being in better nutrition than the geriatric VBC 190 individuals, were not just butchered, defleshed, and marrow-cracked; the spatial patterning of the degraded and broken VBC 202 II.22 bones also indicate having been grease-boiled to extract most protein and fat from the bones themselves after the marrow was extracted. Those bone fragments were then apparently dumped out in the semicircular feature we noted above. The VBC 190 reindeer bones do not indicate grease-boiling, and were simply defleshed, marrow-cracked, and tossed in a circular heap.

This is an interesting difference, because there is obviously more to this story. The VBC 202 II.22 caribou were located proximate to at least three stone-lined hearths, in or just beyond the 20-m^2 secteur. Presumably at least one if not all three of the hearths were actually in use during the human occupation that produced the 202 II.22 faunal assemblage. There were no hearths in or even near VBC Secteur 190. So we have yet another index of human investment beyond the labor and time needed to manufacture stone hearths: the humans in occupation of VBC 202 II.22 invested much more time and labor into processing the caribou than those of VBC 190. It would appear that the geriatric reindeer from VBC 190 were simply unworthy of much investment in labor or time to unlock nutrients from grease-boiling the bones; they weren't smashed into appropriate sizes or shapes to boil. Why do there appear to be two different animal populations at VBC 190 and VBC 2020 II.22? This presumably relates to snapshots in time taken during different seasons. The VBC 202 II.22 fauna indicates a Fall/Winter kill with fawns, yearlings, females, young males, and geriatrics present. VBC 190 fauna suggest a Spring kill, with neonates, younger females, and numerous geriatrics present, while the prime-age males foraged apart from the rest of the herd.

So we should consider the wonderful preservation at VBC and see the various sorts of differential patterning, some of which is radar-visible, as actual material signatures of different human behaviors at the site. And if this is present at VBC it might be present at other sites, in similar AGP-sensible material patterns that can indicate reconstructible aspects of the human past. Simply discounting the shapes and scales of buried material phenomena can be seen thereby as rather unwise and unproductive thinking if not wasteful and neglectful of the archaeological record. There is no reason to assume that only GPR would work at Verberie, neither that any one particular AGP method is "best." This is only one example of just how powerful AGP data are when combined with technical analyses and one's intuition.

CHAPTER 9

Just for Students

Let's pretend that you are seriously interested in archaeological geophysics (AGP) and would like to become involved in it. In that event, you actually have quite a few options available from which to choose. First, you should know that the main limit is the one you place on your own imagination and drive. The history of Archaeology is replete with examples of determined, sometimes lucky, and generally really smart people who pressed through all sorts of obstacles and managed to make a difference in this field for the betterment of all. If that's your goal, you can probably do it too; provided you make all the right moves (it doesn't hurt to get lucky at all the right times), stay out of trouble, and focus on Archaeology, you will have a *chance*. I hate to write this, but Archaeology, like all of American academia, has largely prostrated itself before the debased twin altars of pop-culture celebrity and endless self-promotion on the one hand, and reckless careerism corporatism and bottom-line box-thinking on the other. You might get supremely lucky—as I did—and find mentors who invest time, patience, effort, energy, and money in you because those are the right things to do within the circumstances. But don't bet on it. Bet on YOU and your imagination and intelligence.

- *Disclaimer*: all of the following is premised upon the assumption that even if you want to get away from college right now you will probably have an itch to come back to the Archaeological fold. Those wishing permanently to leave humanistic Anthropological Archaeology forever should make note of that and take what is written below with some serious salt. Despite everything that is wrong with Archaeology, this is still a noble profession (Flannery, 1982), and I offer the following altruistically and in good-faith.

If you really wish to pursue AGP after your baccalaureate graduation but are sick of the grind and don't wish to contemplate grad school, then let's talk a bit about what to do *while you are still in college*. Take some geology, geomorphology, and geophysics courses

Ephemeral Hunter-Gatherer Archaeological Sites. DOI: http://dx.doi.org/10.1016/B978-0-12-804442-1.00009-3

to get a perspective for the physical contexts in which archaeological material resides. Those are marketable things beyond the gown. Don't just enroll in every single Anthro class because it is there. Check with the geoscience or geology department at your school and see if you can work—volunteering if you must—with geophysics scholars and graduate students on their various projects. Everything you do to expand your range of knowledge while you are in school will benefit you later, and the new circles of people you encounter might set you onto new opportunities of which you would otherwise never learn. Moreover, there are probably geophysics jobs you can do right after graduation. I would recommend looking for employment in the construction and/or demolition and also the cable/telecom or even utility industries, especially now that subsurface utility location and concrete-scanning are new and big in those venues. Practically anyone can learn to use geophysical equipment and techniques in such ways, and many employers will pay to have you professionally trained, stake you with good equipment, and probably an enticing salary and perquisites if you are lucky and pass muster. It isn't AGP but it's very similar, and the ideas you will gain from using such stuff outside the confines and pressure of academics will behoove you later in time once you get the academic itch again.

If you *do* wish to pursue AGP as a grad student in an advanced capacity, it's important that you realize you don't have to go specifically into a dedicated geophysics graduate program to gain expertise in AGP (nearly anyone can learn how to use this stuff and remember what I told the undergrads: you can find ways to get involved in this if you look around campus or town). I have several reasons for stating that: First, if you are interested in AGP you should learn about AGP in an environment in which your Archaeology interests are not subordinated by the demands of geophysics. One of my primary criticisms of the scholarship produced by the first wave of archaeological geophysicists—or geophysical archaeologists, as they may be—is this: They write and speak in Geophysics-ese like the attorneys still cling to Latin and French and Olde English phrases in jurisprudence; like physicians still recite Greek or Latin words at you first, and then later explain what they mean. And they have exactly the same motivations. It's quite ironic: in the past 20 years or so AGP technology had advanced to where fortunate people with access to the gear and software (or personal connections to those with them) sallied forth during an AGP Gold Rush to meet their AGP goals and publish everything they could think of—but the published literature clearly indicates that people themselves, human behavior, Anthropology, and even Archaeology were subordinate concerns to those of demonstrating technical virtuosity before self-selected peers in thick physics description and the analysis of really large surface and buried things. They were just beginning to explore the technologies applied to the human past. But they were not and are not Archaeological Paganini.

If you go to grad school to do AGP, keep your mind open; I know it seems like there are Lawgivers who make the Laws and who carefully control AGP reality but there really aren't. There are Gatekeepers, a sort of informal battery of "name-brand" published but "anonymous" AGP scholars who are routinely consulted by the leading journals to review publication submissions and by funding agencies to peruse research proposals, but they haven't thought of everything—I assure you they haven't. There are literally millions of smaller-scale, possible, and worthy research projects and potential publications nobody has even thought of yet. Let those possibilities be your inspirations; you be the ones to tell us about them! Don't let people dissuade you and don't take no for an answer.

One can get a good AGP apparatus for probably less than one would pay for a good used car these days. There are a number of other possibilities that present inside that scenario. For one thing, wind turbines are going to become more common over time, not less, and the CRM work related to them is considerable, and for the wise, potentially lucrative. A personal AGP user could specialize in any number of CRM functions at the expense of pricier, bigger firms using AGP. You could even learn how to specialize within certain particularities of available budget and the going rates.

9.1 COMPETITIVE INTELLIGENCE

We live in a finite world, basically a closed system for our human purposes. In the United States, we also inhabit a very precarious and teetering national economy. Business interests have long known, as have governments since their inception thousands of years ago, that reconnaissance of actual and potential competitors and their activities is useful for analyzing costs and benefits and especially *risks*. A risk is just the probability of sustaining a loss or an injury (somatic, monetary, etc.). Situations that pose the likelihood of risks are known as *perils*. There is an entire minute literature devoted to the theoretical discussion of risks and perils in the insurance industry, and I will avoid indulgence in it here. You are advised to inform yourself about it, however, as well as at least a rudimentary understanding of competitive intelligence. As a student, especially graduate students, you are being subjected to conscious and unconscious competitive intelligence whether you realize it or not. Moreover, the agents of this competitive surveillance are people you interact with probably nearly every day.

Unfortunately for all of us, the contemporary academic policy of insanely competitive "publish or perish" leads to some vitally interesting and potentially unwise behavior. One is the sort of shameless self-promotion to which we are largely inured in the United States, the kind of histrionic hand-waving, attention-seeking that many describe—but we can all plainly see—as the Kardashian effect: the gratuitous attempt to get attention and social currency for its own sake, by virtually any means. In joining web communities, or any community for that matter, one is expressing interest in what they are doing. Having become Facebook friends with a person or group, you are now being surveilled to varying degrees. The things you say and post are noted, maybe replied, but in any event your stuff is now there for others to see. Innocent enough. But what're the grown-up versions of Facebook? Well. . . .

One of them is academic peer review. In writing and submitting original material for publication, you have identified yourself to the publishing venue—and more importantly, to its anonymous professional reviewers—as an interested party, someone who is interested in elevating their scholarly profile, and a potential competitor. You are now, in this economy and the academy in general, a potential *peril* for the established insiders and present them with *risks*. Academic journals all have websites (at least nearly so), upon which are published the names of their elected editorial and advisory boards, people who have vested professional interests (personal prestige, scholarly profile, academic reputation, money, etc.) not only in the journal but also in the scholarly literature in general. The job is presumably to advance science and learning within the spirit of the journal's foundational statements, stated purposes, and among its readership, and to police the literature against perceived pseudoscience or antiscience.

For your information, some of your peers in job searches will pay to obtain competitive intelligence about you in various ways that I will not describe here. Just know that it is happening, it is utterly wicked and commonplace, and there is basically nothing you can do about it except roll up your sleeves and do your best. Just make sure your best is superior to their worst, because not all college students are financially burdened. Many are extremely wealthy and looking for ways to reach out and influence things because that's what they like to do and can do. Professorism tends to occur along family lines, with professors raising other professors, and nurturing their progeny with special insider advantages that you and I simply lack. So your ideas will have to be better and you will need to outwork them since you cannot out-connect them. Anthropology and Archaeology exhibit all of the social pathologies we can see in the larger outside society; they are just magnified and amplified. You will find that it is easier for those folks to slip into opportunities you never even knew existed; some opportunities will be handmade for such people out of thin air and denied you, because you don't have the family ties or the associated "professional" *coterie* that opens doors for them. You're not in the club. Let that sink in before you continue. So please just learn as much as you can, do your level best for yourself and in the spirit of all the investments of love, advice, money, time that your loved ones, friends, and mentors have given you, and, if you really want to do this don't quit unless your health or sanity just can't stand the pressure you place on yourself or the cold, hard stone of the Establishment's disregard is just too much for you to bear.

Anthropology, Archaeology, and AGP have ways of making themselves sound grand and inviting and widely inclusive. That is primarily fluff for public consumption and undergraduate students. Enthroned academics have proprietary interests in their own students (at least most of them), since their students can be expected to propagate many of their mentors' ideas in research and publication, thereby feeding prestige and disciplinary currency back on the mentors. So expect your odyssey to be bumpy when you might submit a grant proposal or a publication manuscript and receive the acidic reviewer comments; you have touched nerves, maybe stepping into the reviewer's comfy niche or a future one, or one they had planned to fill with one of their students. Thing is, peer reviews should never be acidic, sarcastic, desultory, or *ad hominem*; they will be sometimes but they shouldn't be.

Journal editors have to control that process, and it is far easier to just pass comments back and forth than it is to stamp out unprofessional peer reviews. But if you get one, fight back. You can demand that the journal choose another reviewer and see how that goes. You should in any case complain loudly about *ad hominem* attacks in reviews and push back against them. Those are insecure scholars trying to protect their own niches by discouraging you, and this has no place in academic scholarship. So don't tolerate it! If you don't get any critique then you are probably echoing someone's dearly valued own ideas; that is also the death knell of academic scholarship, because if all we do is repeat one another we are regressing. So try to make waves. It's more fun and it's scientifically well-demonstrated to elicit improved thinking. Just make some waves.

All my very best,

JRT

CHAPTER *10*

Questions to Guide Framing Theoretical Perspectives

Included here are suggestions and examples of questions we might wish to ask and attempt to answer in order to frame or reframe AGP research for anthropological utility. In order to avoid the promotion of formulaic, "checklist thinking" the questions lack formal structure.

How can we develop a theory of questioning for Anthropology, Archaeology, and archaeological geophysics (AGP)?

How can we use a theory of questioning to ask better questions?

What is a realistic nonprospecting anthropological objective, or set of objectives, for AGP?

Why are so few archaeologists, whether AAs or GAs, asking questions like this?

Is the entrenched prospecting role for AGP fixed and unshakeable?

If the entrenched prospecting role for AGP fixed and unshakeable, is this because prospecting is more fun and less work?

What is the range of variation archaeological features can be expected to exhibit?

What is the range of variation archaeological features can be expected to exhibit in AGP imagery?

What is the range of variation archaeological features can be expected to exhibit in AGP imagery from SSS?

How can archaeological Feature X be made?

Of what can archaeological Feature X be made?

Where might we expect to find Feature X? In what condition might we find it?

Ephemeral Hunter-Gatherer Archaeological Sites. DOI: http://dx.doi.org/10.1016/B978-0-12-804442-1.00010-X

How might Feature X have been used?

What kinds of human behaviors leave material signatures (MS)?

What would pedestrian survey tell us about Culture or cultures lacking integration with other archaeological techniques?

What would AGP survey tell us about Culture or cultures lacking integration with other archaeological techniques?

How can AGP be used to characterize archaeological material at small-scale site (SSS) remotely?

How can AGP characterization of material at SSS be refined?

What should a comprehensive comparative collection of AGP images include?

Is the literature of contemporary AGP publishing useful for many or most AAs?

Does the literature of contemporary AGP publishing attack theoretical questions or methodological inquiries?

Are the questions posed by contemporary AGP publications based upon Anthropology or another discipline?

Are the questions posed by contemporary AGP publications "meta-disciplinary"?

Should AGP be Archaeology for geophysical use or geophysics for archaeological sake?

What geophysics knowledge should AAs have to use AGP effectively, without misusing it? Who decides?

What archaeological knowledge should GAs have to use AGP effectively, without misusing it? Who decides?

Would greater investment of time and effort in laboratory or sandbox studies devoted to imaging small-scale materials help AGP in general?

Why are there so few published AGP applications at SSS?

Why are there so many published AGP applications at LSS?

How can we forge a better communicative dialogue between AAs and GAs?

Humans have existed on Earth for over 3 million years (3.3 million as I recall), only the past 10,000 of which feature much in the way of megasites: 10,000 is 0.0030303030303030303% of 3.3 million. So what kinds of archaeological sites are probably commonest? Where should we focus most of our AGP attention if we wish to study the human past?

How hard are you willing to work in order to answer your questions, knowing that experts are unlikely to participate or support you?

Can you find AGP equipment to rent/buy so you can use it on your own time and fashion productive simulation research agendas that you can do on your property? Can you build a sandbox?

What is the limit of your imagination?

For ground-penetrating radar: Are there characteristic differences in the dielectrics of limestone hearths? Quartzite hearths? Basalt hearths? Has anyone checked?

For resistivity: How do Paleolithic hunter-gatherer campsites with hearths and middens look in resistivity data?

How do Paleolithic hunter-gatherer campsites with hearths and middens look in magnetic data?

CHAPTER *11*

Conclusion: We Versus Me

This book has made a big deal about experimental archaeological geophysics (AGP) "replication" research and asking questions. The reason for this is mainly that we have no idea what a "representative" small-scale site (SSS) material phenomenon is and we still have a lot to learn about AGP in archaeology. Many questions still remain that don't even actually involve AGP: What is an average SSS? What is an average hearth? What is an average midden? What does an average trash pit look like and how large is it? If we unwisely rule out the AGP use of physical dimensions and scale (Johnson and Haley, 2006), how shall we characterize material phenomenon? What should we analyze? What common elements do such features have? How does an average lithic scatter look in gpr images, or magnetometry data? To even address such questions, we should know the widest possible range of variation in hearth shapes, sizes, and composition. It would also help to know how such things look in a very wide range of AGP data. We have much better ideas about how buried cities, villas, roads, walls, etc., look because such large structures are what the AGP community habitually publishes, to exhaustion.

Large sites, large structures, large phenomena, large pricetags; this is the basis of much contemporary AGP. Few in this community discuss material signatures or human behavior, despite being included in Archaeology and thereby Anthropology (an ethnography of the AGP community would perhaps be very revealing). Behavior is difficult to study, and it might not produce expected results or elicit expected rewards. Human behavior isn't simple and there are no simple material signatures of it (Kelly, 1995). Interpretation of behavioral conditioning in material remains has to allow and to account for wide behavioral variability. Many different behaviors can result in similar material configurations. Likewise, similar behaviors can also result in different material phenomena. Since we know seasonality was a strong influence on past human subsistence, we can understand how even the same human groups could use the same materials or even

Ephemeral Hunter-Gatherer Archaeological Sites. DOI: http://dx.doi.org/10.1016/B978-0-12-804442-1.00011-1

the same sites in very different ways at different times of the year. Moreover, different peer cohorts of the same foraging society could differentially condition familiar materials in very different ways. Site-function and seasonality bear very strongly on material conditioning in SSS (Thompson, 2014, 2015), and such phenomena are virtually unexplored in the AGP literature.

Variability at SSS is a rule and the only way to begin to grasp that variance is to test it experimentally and replicatively, in a wide variety of material configurations, sediment matrices, scales, associations, and with various AGP techniques. It is difficult to envision another effective methodology to address wide behavioral variability and its numerous material effects except through experimental replication and AGP testing. This is how zooarchaeology and lithic technology through replication studies evolved. Unfortunately, everyone wants to advance professionally and especially financially and AGP is still treated as a potential Gold Rush, so many will simply ignore this and earlier (Thompson, 2014, 2015) calls for dedicated ongoing experimental and open-source AGP and the sorts of simulation training I have advocated. Yet, to move away from opportunistic AGP applications for personal benefit, toward an inclusive and really interdisciplinary AGP with anthropological relevance, a dedicated and shared sense of opportunity in mutual discovery should be a paradigm. When I was a student, my peers and I were informed that Anthropology is an holistic discipline devoted to the study and understanding of humankind, our diversity, our evolution, and our shared fates on this planet. Language like that is what hooked me.

Professional responsibility informs us that archaeological materials do not belong to us: they belong to humanity. As Archaeology, and as Anthropology, AGP should not be just another bureaucratic mechanism for self-promotion and self-aggrandizement, as a way to pervert Archaeology into a studied avoidance of manual labor on the one hand (Flannery, 1982) and into a "professionalized" lucrative Establishment for a small set of insiders to exploit on the other. For example, there are, hopefully, numerous graduate and undergraduate students who might enjoy participating in such experimental "ethnoarchaeological" AGP simulations as I describe, using exciting technology and computer software right at the cutting edges of multiple interacting sciences. There is adequate ground space and lab space available, and cheap temporary structures could be erected over some sandboxes and experimental replication "sites" and then folded up and stored at project's end. There would be also plentiful ways to reward enterprising students with class credit, hands-on experience, and other tangible benefits.

A professionally responsible (as in responsible to Anthropology and Archaeology and humanity) AGP community, armed with noninvasive and sophisticated equipment should want an involved experimental replication program, if only for what it could offer all of us. But that would mean subordinating personal expectations of personal gain for disciplinary and interdisciplinary benefit. Given the increasingly "decentralized" *zeitgeist* of our times (i.e., Anonymous, open-sourcing, etc., and see also the 2016 US general election campaign and the decay of political party affiliation) maybe Anthropology (and especially AGP) can participate by decentralizing its careerist properties and encourage the sorts of egalitarian activity it frequently praises in ethnologies and anecdotally. How many of us were exposed to mentors in Anthropology and Archaeology who personally advocated political egalitarianism in unguarded conversations and more intimate extracurricular settings? Was that all just talk? If such principles have real disciplinary meaning, then perhaps we can appeal to them mutually to evolve a more broadly participatory and less mercenary, careerist paradigm for AGP, especially a dedicated program of experimental ethnoarchaeological AGP.

The Tragedy of the Commons (Hardin, 1968) doesn't only describe public landscape destruction, overgrazing, or the decay of road and bridge infrastructure. It also occurs in Anthropology and Archaeology. When Me is more important than We, all of us lose. Partially, this is an artifact of the entrenched "publish or perish" attitude in academics, which is oddly incongruent with much of Anthropology in general. How do anthropologists describe themselves? How, e.g., do we regard the political art of the Persian or Roman Empire as we instruct our students—much of which consists of enormous human figures elevated over miniscule masses? What thematic motifs do we analyze in the Narmer Palette, for another example, as we explain graphic metaphors of power and authority? And then how do we grant-seek, and in what self-promotional terms do we try to one-up the other folks? What do we write about ourselves and how do we see ourselves?

Do we reflect often about our behavior as we seek to monopolize or oligopolize our research projects, their funding, and stake various personal claims inside a "professionalized" bureaucratic field? Any Establishment within Anthropology, Archaeology, and AGP that imagines itself as being a little elite should reflect: it should reflect on υβρις (hubris), and what the word really means. Consider what is discipline-wide alleged to be "tighter research funding" in the context of archaeological research; but not all funding streams have shrunk. The state of the AGP state, e.g., suggests that many Establishment AGP scholars and institutions are funded at a prodigious rate, and the resulting professional literature the money has generated bears visible signs of self-interest and defensive maintenance. An unwise elitist attitude is visible, e.g., in regards to a human prerogative—knowledge and methods of inquiry into our shared human past through Archaeology—and a form of gratuitous, vigorous self-aggrandizement in AGP use and abuse, monopolization and oligopolization in research funding and publication peer reviews. Consider how we present ourselves via social media: are our sites more like Conyers': open, informative, and accessible to anyone with interest? Or are they mere self-promotion devices ultimately premised upon financial gain or elbowing out perceived competitors trying to reach into *cosa nostra*?

In one sense, this and earlier calls for a more collegial and mutualized AGP are case studies in the Anthropology of Contemporary Anthropology, Archaeology, and AGP. Receptivity to, resistance to, and also ignorance of them represent real data—data about whether Anthropology at large, but specifically Archaeology, believes the things it says about itself and about humanity. This is information about whether AGP can evolve into a humanistic enterprise or is just another bureaucracy, intended and managed to channel resources in specific directions. In an earlier work, I distributed a questionnaire to some archaeologists, including AGP specialists, CRM professionals, academic archaeologists, and a few geologists. The abysmal rate of informant participation convinced me that a future more representative sampling was impossible because most requests would be simply ignored by colleagues. Yet, the few who did participate related themes that should alarm the AGP Establishment (Thompson, 2015). The AGP community, such as it is, should also reflect on the relative lack of popularity "interest groups" enjoy today, mainly due to how they attempt to manage affairs in their own self-image and self-interest.

There are sound cultural justifications for the contempt and ill-repute in which our larger communities hold narcissism and self-aggrandizement amongst self-interested subsets.

Thompson (2015) indicates several troubling themes: among them are an attitude among CRM employees that AGP will put diggers out of business and so some actively resent it; that AGP surveys are too costly and too iffy for widespread CRM application; that among academic anthropological archaeologists a general lament exists that their input in AGP scholarly literature is unwanted, and is in fact regarded as turf-intrusion by an established elite and is thereby punitively peer reviewed to exclude them; that AGP is selfishly and narcissistically devoted to exploration of theoretical premises that are irrelevant to Anthropology and Archaeology in general, and its scholarly literature is policed by insiders against encroachment from without. While this is not a call to open the gates to the barbarians, it is a request to get the AGP Establishment to reflect and reconsider its role now and its potential to participate in a future from which we could all benefit. For if the general public can teach themselves to use metal-detectors to find buried metallic valuables in the ground, this is also another signal to AGP and its self-interested cohort. People can and will learn how to use and "do" AGP, whether you like it or not and whether you control it or not. The training courses I received at GSSI, e.g., suggest that nearly anyone can learn to use AGP equipment in extremely sensitive and sophisticated ways. The obvious self-interest that manufacturers have in broadening their base of consumers and users is likewise important data for consideration by the AGP Establishment.

Why is AGP so pervasively regarded as "prospection" by its Establishment instead of observation? Why are AGP applications called "surveys" and not tests? Why is there such a continual literary push to stake method claims on the competitive use of Technique X over here because it worked over there? Are those signs that AGP is "working"? For whom does this work? What more could it be? Who is left out? Why aren't more people asking these questions? The noninvasive power of AGP offers potentially tremendous benefit to all archaeologists. That more American academic and CRM archaeologists don't consider AGP as a routine constituent activity is therefore also valuable data. Why don't more American archaeologists see AGP as basically a "given" aspect of Phase I or Phase II activities? If most

American Anthropological, academic, and CRM archaeologists are generally ignorant of AGP methods (Conyers, 2013), abilities, or potential benefits *why* is that the case? Has AGP been presented in such ways as to engage, invite, encourage, enlist, and support the active cooperation of nongeophysicists in research, publishing, application, method, and demeanor?

One suspects that many CRM professionals are actually acutely aware of AGP—and the ways in which it is presented to and inflicted upon the lesser firms by larger firms and others with access to the equipment: as an explicitly for-profit venture by a small group of profiteers. Small CRM firms are used to counting beans, and they're probably much more attuned to nuts and bolts costs and expenses than most. I have personally witnessed multiple instances of such monetized self-interest applied in the realm of CRM; such occurrences are especially infuriating when they involve time-sensitive data-withholding and refusal to distribute data pending payment. Cash and carry ought not to be a policy among Anthropologists. These are not proud legacies of AGP, at least not in the United States, but perhaps they shouldn't be surprising. In many respects, broader involvement of CRM firms with AGP should also be viewed skeptically, since there will be attempts made to privatize, monetize, and monopolize even further. Thus larger CRM firms with budgets adequate to purchase AGP equipment and to conduct AGP applications for smaller firms—for hire, of course—as private subcontractors should also do some self-reflection. Just because you can does that mean you should? I am a big believer in the ability of humanity to solve its mutual problems by magnifying We and minimizing Me; this got us through the Pleistocene, e.g.; the collective behavior of risk-pooling, cost-sharing, and altruistic behavior. We need more We and less Me.

This volume is also not a lecture to chide unduly human mortals for exhibiting bad behaviors; goodness knows I have more than a few of them myself. We all suffer the curse of acquisitiveness and selfishness, to varying degrees and intensities. But it is a call to remind people who are first and foremost anthropologists by education, training, and expertise to remember that Anthropology in the United States is about much more than dollars. In an Anthropological sense, would the general public describe AGP and its regrettable bias toward thing-finding as just another form of treasure-hunting? Furthermore, there are some extremely disquieting possibilities represented by private and even personal nonspecialists using AGP equipment for pot-digging and site-mining activities for the archaeology black market. Academic and CRM archaeologists are well aware of such dangers to our shared cultural resources, and these themes should be part of AGP training in Anthropology departments. Does the AGP Establishment conduct treasure-hunting, with digital data as the treasure and knowledge of application as the hunt? Are those humanistic or even Anthropological, to say nothing of noble, goals? If AGP is just another method, just another technique, then nonspecialists can and will learn how to use it and to bend it to their own purposes. Isn't that what many AGP applications represent today, small profit-generating opportunities, even when conducted by "professionals"? Is, therefore, a monopolistic or oligopolistic model for AGP professionally responsible? If, as Conyers (2012, 2013) suggests, Archaeologists lacking training in AGP need to embrace the AGP Establishment, should we be certain that Archaeologists will be embraced in return? Does the AGP Establishment welcome or despise input and research from Anthropological Archaeologists? Untrained pot-diggers could follow using AGP, and probably already are. Should not a broader active coalition of AGP initiates better protect our fragile remaining cultural resources? As a final and simple note, I suggest that all of us who are interested in using AGP could do with a lot more We and drastically less Me.

REFERENCES

Audouze, 2016. Personal communication.

Binford, L., 1978. *Nunamiut Ethnoarchaeology*. Academic Press.

Booth, A., et al., 2015. Structure of an ancient egyptian tomb inferred from ground-penetrating radar imaging of deflected overburden horizons. Archaeol. Prospect. 22 (1), 33–44.

Comins, N., Kaufmann, W., 2005. Discovering the Universe, eighth ed Freeman and Company.

Conyers, L., 2010. Ground-penetrating radar for anthropological research. Antiquity 84 (323), 175–184.

Conyers, L., 2012. Interpreting Ground-Penetrating Radar for Archaeology. Left Coast Press.

Conyers, L., 2013. Ground-Penetrating Radar for Archaeology, third ed AltaMira Press.

Conyers, L., Leckebusch, J., 2010. Geophysical archaeology research agendas for the future: some groundpenetrating radar examples. Archaeol. Prospect. 17, 117–123.

Creekmore, A., 2010. The structure of upper mesopotamian cities: insight from fluxgate gradiometer survey at Kazane Hoyuk, Southeastern Turkey. Archaeol. Prospect. 17 (2), 73–88.

Enloe, J., 2004. Equifinality, assemblage integrity and behavioral inferences at Verberie. J. Taphonomy 2 (3), 147–165.

Enloe, J., 2006. Geological processes and site structure: assessing integrity at a late paleolithic open-air site in Northern France. Geoarchaeology 21 (6), 523–540.

Enloe, J., 2010. Technology and demographics: an introduction. In: Zubrow, E., Audouze, F., Enloe, J. (Eds.), The Magdalenian Household: Unraveling Domesticity. SUNY Press, pp. 11–14.

Flannery, K., 1982. The golden Marshalltown: a parable for the archeology of the 1980s. Am. Anthropol., New Series 84 (2), 265–278.

Gibbon, G., 2014. Critically Reading the Theory and Methods of Archaeology: An Introductory Guide. AltaMira Press.

Hardin, G., 1968. The tragedy of the Commons. Science, New Series 162 (3859), 1243–1248.

Jaubert, J., et al., 2016. Early Neanderthal constructions deep in Bruniquel Cave in southwestern France. Nature. Available from: http://dx.doi.org/10.1038/nature18291.

Johnson, J. (Ed.), 2006. Remote Sensing in Archaeology: An Explicitly North American Perspective. University of Alabama Press.

Johnson, J., Haley, B., 2006. A cost-benefit analysis of remote sensing application in cultural resource management archaeology. In: Johnson, J. (Ed.), Remote Sensing in Archaeology: An Explicitly North American Perspective. University of Alabama Press, pp. 33–46.

Kelly, R., 1995. The Foraging Spectrum: Diversity in Hunter-Gatherer Lifeways. Smithsonian Institution Press.

Kvamme, K., 2003. Geophysical surveys as landscape archaeology. Am. Antiq. 68 (3), 435–457.

Leckebusch, J., Peikert, R., 2001. Investigating the true resolution and capabilities of ground-penetrating radar in archaeological surveys: measurements in a sand box. Archaeol. Prospect. 8, 29–40.

Mallet, J., 2008. Hybridization, ecological races and the nature of species: empirical evidence for the ease of speciation. Philos. Trans. R. Soc. B 363, 2971–2986.

Neubauer, W., et al., 2002. Georadar in the Roman civil town Carnuntum, Austria: an approach for archaeological interpretation of GPR data. Archaeol. Prospect. 9 (3), 135–156.

Rogers, M., et al., 2010. Cesium magnetometer surveys at a Pithouse Site near Silver City, New Mexico. J. Archaeol. Sci. 37, 1102–1109.

Sagan, C., 1995. The Demon-Haunted World: Science as a Candle in the Dark. Random House.

Sankararaman, S., et al., 2014. The genomic landscape of Neanderthal ancestry in present-day humans. Nature. Available from: http://dx.doi.org/10.1038/nature12961.

Schneider, A., et al., 2015. A template-matching approach combining morphometric variables for automated mapping of charcoal kiln sites. Archaeol. Prospect. 22 (1), 45–62.

Solla, M., et al., 2012. Experimental forensic scenes for the characterization of ground-penetrating radar wave response. Forensic Sci. Int. 220 (1–13), 1–9.

Storey, 1997. Personal communication.

Testone, V., et al., 2015. Use of integrated geophysical methods to investigate a coastal archaeological site: the Sant'Imbenia Roman Villa (Northern Sardinia, Italy). Archaeol. Prospect. 22 (1), 63–74.

Thompson, J., 2011. Views to the Past: Faunal and Geophysical Analysis of the Open-Air Upper Paleolithic Site of Verberie. Unpublished Doctoral Dissertation on file at The University of Iowa and on ProQuest and UMI.

Thompson, J., 2012. The allegory of the hammer and the nail gun and other unstable orthodoxies of 'modernity': possible pitfalls of 'behavioural modernity'. The Newsletter of the Australian Rock art Research Association 29 (2), 3–12.

Thompson, J., 2014. A Composite View to the Past: Integrating Zooarchaeology and Archaeological Geophysics Methodologically at the Magdalenian Site of Verberie le Buisson-Campin. British Archaeological Reviews/Archaeopress.

Thompson, J., 2015. Anthropological Research Frames for Archaeological Geophysics: Material Signatures of Past Human Behavior. Rowman & Littlefield.

Thompson, V.D., Arnold III, P.J., Thomas, J., Pluckhahn, VanDerwarker, A., 2011. Situating remote sensing in anthropological archaeology. Archaeol. Prospect. 18 (3), 195–213.

Vernot, B., Akey, J., 2014. Resurrecting surviving neandertal lineages from modern human genomes. Science. Available from: http://dx.doi.org/10.1126/science.1245938.

Printed in the United States
By Bookmasters